CONTROL AND DYNAMIC SYSTEMS

Advances in Theory and Applications

Volume 13

CONTRIBUTORS TO THIS VOLUME

G. S. CHRISTENSEN
M. E. EL-HAWARY
T. T. FENG
RANDALL V. GRESSANG
CHRISTIAN GUENTHER
MICHAEL D. INTRILIGATOR
E. J. HAUG
VICTOR LARSON
PETER W. LIKINS
A. N. MICHEL
R. D. RASMUSSEN
JOHN F. YOCUM, JR.
DEMETRIUS ZONARS

CONTROL AND DYNAMIC SYSTEMS

ADVANCES IN THEORY AND APPLICATIONS

Edited by
C. T. LEONDES

SCHOOL OF ENGINEERING AND APPLIED SCIENCE
UNIVERSITY OF CALIFORNIA
LOS ANGELES, CALIFORNIA

VOLUME 13 1977

ACADEMIC PRESS New York San Francisco London
A Subsidiary of Harcourt Brace Jovanovich, Publishers

ACADEMIC PRESS RAPID MANUSCRIPT REPRODUCTION

ACADEMIC PRESS, INC.
111 Fifth Avenue, New York, New York 10003

United Kingdom Edition published by
ACADEMIC PRESS, INC. (LONDON) LTD.
24/28 Oval Road, London NW1

LIBRARY OF CONGRESS CATALOG CARD NUMBER: 64-8027

ISBN number: 0-12-012713-X

PRINTED IN THE UNITED STATES OF AMERICA

CONTENTS

CONTRIBUTORS

Numbers in parentheses indicate the pages on which the authors' contributions begin.

G. S. Christensen (1), Department of Electrical Engineering, University of Alberta, Edmonton, Alberta, Canada

M. E. El-Hawary (1), Faculty of Engineering and Applied Science, Memorial University of Newfoundland, St. John's, Newfoundland, Canada

T. T. Feng (207), Department of Mechanics, University of Iowa, Iowa City, Iowa

Randall V. Gressang (161), Flight Control Division, Air Force Flight Dynamics Laboratory, Wright-Patterson Air Force Base, Ohio

Christian Guenther (71), Messerschmitt-Bolkow-Blohm, Munich, West Germany

E. J. Haug (207), Concepts and Technology, AMSAR/RDT, U.S. Army Armaments Command, Rock Island, Illinois

Michael D. Intriligator (135), Economics Department, University of California, Los Angeles, California

Victor Larson (285), Jet Propulsion Laboratory, Pasadena, California

Peter W. Likins (285), School of Engineering and Applied Science, University of California, Los Angeles, California

A. N. Michel (323), Department of Electrical Engineering and Engineering Research Institute, Iowa State University, Ames, Iowa

R. D. Rasmussen (323), Department of Electrical Engineering and Engineering Research Institute, Iowa State University, Ames, Iowa

John F. Yocum, Jr. (247), Space and Communications Group, Hughes Aircraft Company, El Segundo, California

Demetrius Zonars (161), Air Force Flight Dynamics Laboratory, Wright-Patterson Air Force Base, Ohio

PREFACE

The theme for this volume is the *Techniques of Control and Dynamic Systems and Their Application to Modern Complex Engineering, Industrial, and Other Systems.*

In the past, the volumes in this annual series have consisted of diverse, interesting, and significant contributions in the very broad and rather complex field of control and dynamic systems. This volume marks the second time in this series wherein a timely theme volume of a significant subject area of lasting interest is included.

In the modern era of control and dynamic systems, after about one and one-half to two decades of the development of modern techniques, it is increasingly apparent that major advances are developed through the expedient of application of these techniques to modern complex systems. These applications to modern complex systems are themselves of import, and over time, with the establishment of a broader base of knowledge of these modern techniques on the international scene, a much richer array of substantive applications will quite naturally result. The future is most certainly exciting and important in this regard. The thrust of this volume is in this major broad area.

CONTENTS OF PREVIOUS VOLUMES

Volume 7

Volume 8

Volume 9

Volume 10

OPTIMAL OPERATION OF LARGE SCALE POWER SYSTEMS

M. E. EL-HAWARY

Faculty of Engineering and Applied Science
Memorial University of Newfoundland
St. John's, Newfoundland, Canada

G. S. CHRISTENSEN

Department of Electrical Engineering
University of Alberta
Edmonton, Alberta, Canada

* The work presented in this chapter was supported by the National Research Council of Canada under Grants A-4146, A-9050 and the Brazilian Bank of National Development (BNDE).

I. INTRODUCTION

A prime objective in the operation of a power system is to achieve optimum economic dispatch. This is the problem of scheduling the generation at various plants, such that the power demand is supplied at minimum production cost. The optimum operation of a power system will depend on restrictions imposed by factors other than operating economics. Possible decreases in power production costs are of vital concern to the electric utility industry. In addition, the capability to solve the economic dispatch problem is extremely useful for the planning and design of future equipment additions. For these reasons, the economic dispatch problem has been the subject of extensive research [1, 2].

In economic dispatch it is customary to consider the

operating costs only. This ignores expenses of capital, labor, startup and shutdown related to the length of the outage period for a certain unit. It is essential to have an accurate knowledge of the manner in which the total cost of operation of each available energy source varies with the instantaneous output. The appropriate price to use for economic dispatch is the current cost of incoming fuel adjusted for handling costs, maintenance cost of fuel handling facilities, etc.

The hydrothermal optimization problem is different from the all-thermal one. The former involves the planning of the usage of a limited resource over a period of time. The resource is the water available for hydrogeneration. Most of the hydro-electric plants are multipurpose in nature. In such cases it is necessary to meet certain obligations other than power generation. These may include a maximum forebay elevation not to be exceeded due to flood prospects and a minimum plant discharge and spillage to meet irrigational and navigational commitments. Thus the optimum operation of the hydrothermal system depends on the conditions that exist over the entire optimization interval [3]. One system with large water storage capacity may require a year for the optimization interval; another system may have run-of-the-river plants with only a small or moderate storage capacity. An optimization interval of a day or a week may be useful in this case [4].

Other distinctions among power systems are the number of hydro stations, their location and special operating characteristics. The problem is quite different if the hydro stations are located on the same stream or on different ones. In the former case, the water transport delay may be of great importance [5, 6]. An upstream station will highly influence the operation of the next downstream station. The latter, however, also influences the upstream plant by its effect on the tail water elevation and effective head. Close coupling of stations by such a phenomenon is a complicating factor.

A. A History of Economy Dispatch.

A brief presentation of previous investigations of hydrothermal power systems economic operation will be given. Various optimization techniques have been used in the past. Among these are the variational calculus principles, the methods of dynamic programming, and the Pontryagin's maximum principle.

In 1940, Ricard [7] obtained a set of operating schedules for a hydrothermal system with no losses. His work was continued in 1953 by Chandler *et al.* [8] who included transmission losses but with constant hydraulic head. The latter method was improved in 1958 by Glimn and Kirchmayer [9] who included variable head plants. They also reported the work of Kron, who developed equivalent equations. A set of scheduling equations was developed in 1953 by Cypser [5]. These were developed under the assumption that variations in elevation and plant efficiencies can be neglected. Carey [10] suggested an approach that would linearize Cypser's equations. Watchorn [3] gives a set of equations to be satisfied in order to achieve maximum economy.

The above-mentioned investigations employed the Euler equation of the calculus of variations to obtain the scheduling equations. The work of Arismunander [11] and Arismunander and Noakes [12] dealt with short-range optimization of hydrothermal systems. All the necessary and sufficient conditions for optimality of the variational calculus were employed. In addition the equivalence of all previously developed equations were proved. In 1962, Drake *et al.* [6] presented a dispatch formula based on the calculus of variations. This formula is restricted to the case where all the hydro plants operate with constant head. The system considered has series plants, multiple chains of plants, and intermediate reservoirs. Kirchmayer and Ringlee [13] presented a dispatch formula for a hydrothermal power system in 1964. Head variations were considered. The formula applies for power systems having one hydro plant. Discussing the work of Drake *et al.* [6], Watchorn [3] points out the importance of considering variable head for the optimization of such systems. In separate discussions of the same work, Watchorn and Arismunandar point out that a river time delay of a couple of hours is highly important for accurate optimization of many power systems.

In 1960 Bernholtz and Graham [14] presented a dynamic programming solution to the hydrothermal optimization problem. The application of both the Pontryagin's maximum principle and dynamic programming to the hydrothermal dispatch problem was considered by Dahlin [4] in 1964 and later, jointly with Shen [15]. The general dispatch formulas obtained were applicable to a wide class of systems. These were the systems with hydro plants having fixed head, varying head, and hydraulic coupling, both with and without river transport dealy. The economic operation of a simplified model system and the long range operation of a multireservoir system is considered by Hano *et al.* [16] in 1966. They employed the Pontryagin's maximum principle to obtain the scheduling equations.

The various factors involved in hydrothermal coordination and their interrelation required for optimum generation are discussed by Watchorn [3] in 1967. In a discussion of Watchorn's paper, Christensen points out that the scheduling equations obtained constitute only a necessary condition for optimality. He suggested the steepest descent method to search for the global optimum mode of operation.

The problems considered in this chapter are characterized by the presence of common-flow hydro plants in the system. These problems are of a complex nature. Among the early contributions to this problem is Burr's [17]. He developed loading schedules for a two-plant common-flow hydro system but the assumptions made were too simplifying. Later, Menon [18] used the Euler equations for constructing sets of minimizing sequences for a three-plant hydrothermal system. It is noted here that the system considered by Menon was of low dimension, that is, a small number of hydro plants were considered. Also this was a long-range scheduling problem.

In 1966 Dahlin and Shen [15] treated the problem of a power system with hydro plants on the same stream using the Pontryagin's maximum principle. They used a river flow model which introduced a large number of differential equations and boundary conditions. This was a definite contribution to the theory of economy scheduling. Unfortunately this model made the problem more difficult to analyze numerically.

A more recent related work is that by Miller and Thompson [19]. Their work is concerned with the Pacific Gas and Electric Company hydrothermal system. A linear programming approach is used for solving the long-range scheduling problem. A set of inequality constraints on the reservoir's storage and head variations are imposed. However, the time delays of flows were not taken into consideration.

The extension of the existing economy dispatch solutions to include the exact model of the transmission network is due to Carpentier [20]. The resulting optimization problem was shown to be one of nonlinear programming. Necessary conditions for optimality were derived using the nonlinear programming techniques. In their paper [21], Peschon *et al.* presented the general problem considered by Carpentier for an all-thermal system. Another important contribution is that of Dommel and Tinney [22]. Here optimal power flow solutions are obtained for an all-thermal system. The method is based on power flow solution by Newton's method [23], and a gradient adjustment algorithm for obtaining the minimum is employed.

A unified approach to load flow, minimum loss, and economic dispatching problems was presented by Sasson [24], who investigated the application of various nonlinear programming methods to the problem. In 1969 El-Abiad and Jaimes [25] presented a variational method to solve the optimal load flow problem. It is noted that these two works were also concerned with all-thermal systems.

The problem of power systems reliability motivated the work of Sullivan and Elgerd [26]. An effort to define a reliability objective in terms of the system's reactive power generations was made. The basic idea of their work was to distribute optimally the reactive power generation between the system generators.

The work by Shen and Laughton [27] was of the same nature as those previously mentioned. The main contribution was in exploring the problem of existence and uniqueness of the optimal solution using nonlinear programming techniques. The problem of a hydrothermal system with negligible head variations was solved by Ramamoorthy and Rao [28]. Here a discrete formulation was adopted and the problem was solved using the nonlinear programming techniques. The discretization process makes the problem one of a large dimension. However, a method of splitting the problem into ones of smaller dimension was proposed.

A dual linear programming formulation was given by Shen and Laughton, [29], who obtained a fast solution for the problem of an all-thermal system under inequality constraints. This was a contribution to the on-line dispatching problem. The need for including objectives other than economy is illustrated by Gent and Lamont [30], who considered a minimum emission dispatch problem. Reference is made here to the paper by Sasson and Merrill [31] on the optimal load flow and applications of optimization techniques to power system problems. For a thorough treatment of economy operation of power systems Kirchmayer's book [32] is a main reference.

B. The Functional Analytic Optimization Techniques

1. Historical Survey. During the years optimal control

was being developed, powerful general solution methods were introduced. These are based on the now widely known "maximum principle" and "optimality principle". Parallel to this development, starting in 1956, attempts were being made to introduce methods of functional analysis into the study of optimal control problems.

At first it seemed that the methods of functional analysis applied only to a very restricted class of problems. In spite of this, however, the number of studies using the ideas of functional analysis has increased. In solving optimal control problems, by using the maximum principle, or by reduction to the Euler equations, these methods do not show how to select the initial conditions required for solving the adjoint system. The methods of dynamic programming and the approach that leads to the Hamilton-Jacobi equations do not have this deficiency. However, the solution of functional equations is not an easy problem [33].

One of the typical features of the functional analysis approach is that it yields necessary and sufficient conditions for the existence of solutions. This fact makes it possible to study the qualitative aspects of optimal processes. Moreover, this approach is free of the concrete nature of the system. Thus many formulations hold for systems that are distributive, digital, nonlinear, or biological. Of course, results obtained on the basis of an abstract formulation must then be given concrete identification in its various physical forms.

In the following we give a survey of certian works where the methods of functional analysis are used in solving problems in the theory of optimal processes. It is not the authors' intention to give a complete exposition of the application of functional analysis to the theory of optimal processes. In fact, works dealing with the abstract minimum norm formulation will be our main concern.

Investigation of the problem of approximate solutions to

first-order ordinary differential equations, lead Carter [34], in 1957, to one of the earliest minimum norm formulations. Carter's problem was concerned with obtaining an element of a specific Banach space. The image of the element sought under a first-order linear differential operator was to be of minimum norm. The norm adopted was the maximum norm. The element was to satisfy a two-point boundary condition.

In 1962, Reid [35] extended Carter's results to the case of an nth-order differential operator. This was achieved by reducing the problem to a problem in the theory of moments. The general results of the Hahn-Banach theorem were then applied to the reduced problem.

A minimum norm problem in Hilbert spaces was considered by Balakrishnan [36] in 1963. A wide class of minimum norm problems was considered by Neustadt, [37], who employed a variational approach to find the element with minimum norm. This was set in the Banach spaces of the L_p type $(1 < p < \infty)$. The system satisfied a linear integral operator and the problem was reduced to minimizing a functional of a new variable. This variable is in many ways analogous to the costates of the Pontryagin's maximum principle. The method of steepest descent was suggested for implementing the final optimal.

Kranc and Sarachick [38] considered a minimum norm problem in 1963. This was essentially the same as Neustadt's problem. The only exception was that the element (control) sought was to belong to a specified ball in the L_p space under consideration. Hölder's inequality was used to specify the optimal solution.

In 1964 Porter [39] considered a problem involving a linear transformation on a "Hilbert space". The cost associated with an element of the Hilbert space was given by the Hilbert space norm. Later, together with Williams [40, 41] he extended the results of this abstract problem to cases involving Banach spaces. The results of these approaches are applicable to systems of discrete, continuous, and composite types. These results can be utilized for various optimization problems [42-44]. The 1964 contribution of Hsieh [45] to the synthesis of adaptive control systems involved what we call the minimum norm approach.

The functional analytic optimization technique employed in solving the power system problem will be outlined here. For basic concepts and theoretical development we refer the reader to the literature [46-48].

2. The Minimum Norm Problem. The results reported in this chapter are based on the formulation of an abstract minimum norm problem. This is formulated by Porter [46].

Let B and D be Banach spaces. Let T be a bounded linear transformation defined on B with values in D. For each ξ in the range of T, find an element $u \in B$ that satisfies

$$\xi = Tu \tag{1}$$

while minimizing the performance index

$$J(u) = \|u\|. \tag{2}$$

THEOREM. The unique optimal $u_\xi \in B$ is given by

$$u_\xi = T^\dagger \xi \tag{3}$$

where the pseudoinverse operator $T^\dagger$ is given by

$$T^\dagger \xi = KT^*J^{-1}\xi. \tag{4}$$

In Hilbert spaces the operator K reduces to the identity operator so that

$$T^{\dagger}\xi = T^*[TT^*]^{-1}\xi, \tag{5}$$

provided that the inverse of TT^* exists.

An extension of the results of the minimum norm problem discussed is the following.

Let B, D, T, and ξ be as in the minimum norm problem. Let $\hat{u}$ be a given vector in B. Then the unique $u_\xi \in B$ satisfying

$$\xi = Tu$$

which minimizes the performance index

$$J(u) = \|u-\hat{u}\| \tag{6}$$

is given by

$$u_\xi = T^{\dagger}[\xi - T\hat{u}] + \hat{u}. \tag{7}$$

II. THE POWER SYSTEM MODELS

A. The Electric Network

The system's electric network is represented by N buses (or nodes), which are connected by branches or lines having conductance G^{ij} and admittance B^{ij}. Connected between bus i and neutral is a branch having conductance G^{i0} and admittance B^{i0}. This is required for the equivalent π representation of transmission lines.

At a given bus i, the net active power P_i and reactive power Q_i are given by

$$P_i(t) = E_i^2(t)G_i - E_{d_i}(t) \sum_{\substack{j=1,\\ j\neq i}}^{N} [E_{d_j}(t)G^{ij} - E_{q_j}(t)B^{ij}]$$

$$- E_{q_i}(t) \sum_{\substack{j=1,\\ j\neq i}}^{N} [E_{d_j}(t)B^{ij} + E_{q_j}(t)G^{ij}], \quad i=1,\ldots,N, \qquad (8)$$

$$-Q_i(t) = E_i^2(t)B_i + E_{q_i}(t) \sum_{\substack{j=1,\\ j\neq i}}^{N} [E_{d_j}(t)G^{ij} - E_{q_j}(t)B^{ij}]$$

$$- E_{d_i}(t) \sum_{\substack{j=1,\\ j\neq i}}^{N} [E_{d_j}(t)B^{ij} + E_{q_j}(t)G^{ij}], \quad i=1,\ldots,N, \qquad (9)$$

where $G_i = \sum_{\substack{j=0,\\ j\neq i}}^{N} G^{ij}$, $\qquad B_i = \sum_{\substack{j=0\\ j\neq i}}^{N} B^{ij}$.

Note that if the E_i's are the phasor voltage to neutral, then it is obvious that $E_{d_0} = E_{q_0} = 0$.

Equations (8) and (9) are the load flow equations. Each bus is characterized by four variables $P_i(t)$, $Q_i(t)$, $E_{d_i}(t)$, and $E_{q_i}(t)$. In a normal load flow study, two of the four variables are specified and the others must be found. Depending on which variables are specified, the buses can be divided into three types [49]:

(1) <u>generator bus</u> with P and E specified, Q and the phase angle $[\tan^{-1}(E_q/E_d)]$ unknown,

(2) <u>load bus</u> with P and Q specified, E_d and E_q being the unknowns, and

(3) <u>slack bus</u> with E_d and E_q specified, P and Q unknowns.

For convenience this shall be the node N_g and $E_q(t)$ is taken as zero. Since the slack bus is taken as a generator N_g node, this means that the number of the unknowns is reduced by one. We may assume that E is not specified at the $(N_g - 1)$ bus.

There are several inequality constraints that must be satisfied. These are given by

$$P_i^2(t) + Q_i^2(t) \leq S_i^{2^M}, \qquad i = 1,\ldots,N_g, \tag{10}$$

$$Q_i^m \leq Q_i(t) \leq Q_i^M, \qquad i = 1,\ldots,N_g, \tag{11}$$

$$P_i^m < P_i(t) \leq P_i^M, \qquad i = 1,\ldots,N_g. \tag{12}$$

Furthermore, since E at a generator bus is specified then one requires

$$E_{d_i}^2(t) + E_{q_i}^2(t) = E_i^2(t), \qquad i = 1,\ldots,N_g - 1, \tag{13}$$

where $E_i(t)$ is assumed to be known.

The load flow formulation introduces a large number of nonlinear equations to the economy dispatch problem. For a comprehensive review of load flow solution techniques Stott's [52] paper is recommended. The inclusion of the electric network model in the economy operation problem is traditionally affected by considering only the active power balance.

The inclusion of the system's transmission losses in the power balance equation is based on the well-established loss formula [32]. It is noted that such a function as the transmission losses cannot be expressed in terms of only generator powers in an exact manner. There are a number of approximations involved in the loss formula derivation. However, it produces close answers with errors of up to a few percent [53]. Very sophisticated methods of calculating the loss formula coefficients exist and are being used [54, 55]. The generation schedule sought must satisfy the active power balance equation

$$P_D(t) = \sum_{i=1}^{N_g} P_i(t) - P_L(t). \tag{14}$$

The transmission loss is a quadratic function of the active power generated by the system plants and is given by [32]

$$P_L(t) = \sum_{i=1}^{N_g} \sum_{j=1}^{N_g} P_i(t) B_{ij} P_j(t) + \sum_{i=1}^{N_g} B_{i0} P_i(t) + K_{L0} \tag{15}$$

with B_{ij}'s and B_{i0}'s and K_{L0} being the loss formula coefficients which are assumed to be known, with the property

$$B_{ij} = B_{ji}, \qquad i,j = 1,\ldots,N_g.$$

B. The Hydro System

We will present the case of a power system with series plants (on the same stream), multiple chains of plants, and intermediate reservoirs (Fig. 1). The variety of models that can be considered from a theoretical standpoint is infinite. However, a practical model is chosen in this section. Here the formulation adopted is applicable to any practical system with a larger number of hydro plants.

The ith hydro plant's active power generation $P_{h_i}(t)$ is given by

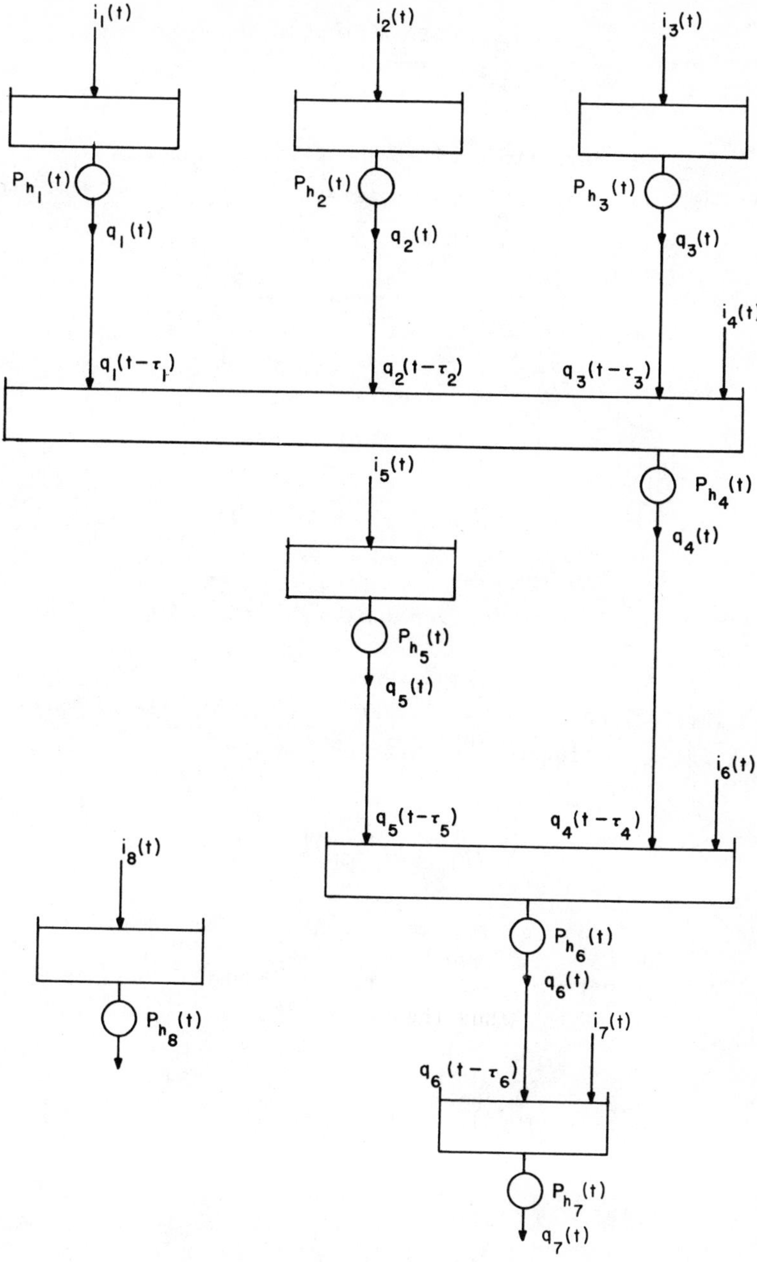

Fig. 1. General layout of the hydro plants for the system

$$P_{h_i}(t) = \frac{n_i q_i(t) h_i(t)}{11.8} \text{ kW}. \tag{16}$$

The effective hydraulic head at the plant is equal to the difference between the forebay elevation $y_i(t)$ and the tail-race elevation $y_{T_i}(t)$, thus

$$h_i(t) = y_i(t) - y_{T_i}(t). \tag{17}$$

The forebay elevation $y_i(t)$ is given by

$$y_i(t) = y_{i0} + \beta_{y_i} s_i(t) \tag{18}$$

This relation is true for vertical sided reservoirs, with y_{i0} and β_{y_i} constants corresponding to the forebay geometry.

The tail-water elevation varies with the rate of water discharge according to the relation

$$y_{T_i}(t) = y_{T_{i0}} + \beta_{T_i} q_i(t), \tag{19}$$

where $y_{T_{i0}}$ and β_{T_i} are known constants corresponding to the tail-race geometry. Thus the effective head $h_i(t)$ is

$$h_i(t) = \alpha_i + \beta_{y_i} s_i(t) - \beta_{T_i} q_i(t), \tag{20}$$

where

$$\alpha_i = y_{i0} - y_{T_{i0}}, \tag{21}$$

so that

$$P_{h_i}(t) = \frac{q_i(t)}{G_i}[\alpha_i + \beta_{y_i} s_i(t) - \beta_{T_i} q_i(t)]. \qquad (22)$$

Here we have $G_i = 11.8/\eta_i$ as an efficiency constant.

For the upstream plants, the reservoir dynamics are given by

$$\dot{S}_i(t) = i_i(t) - q_i(t). \qquad (23)$$

Here $i_i(t)$ is the natural inflow to the reservoir. Define

$$D_i(t) = S_i(0) + \int_0^t i_i(\sigma)\, d\sigma, \qquad (24)$$

$$Q_i(t) = \int_0^t q_i(\sigma)\, d\sigma, \qquad (25)$$

$$A_i(t) = -[\alpha_i + \beta_{y_i} D_i(t)]/G_i, \qquad (26)$$

$$B_i = B_{y_i}/G_i, \qquad (27)$$

$$C_{m+i} = \frac{\beta_{T_{m+i}}}{G_{m+i}}. \qquad (28)$$

The upstream plants' active power generations satisfy

$$P_{h_i}(t) + A_i(t)q_i(t) + B_i q_i(t)Q_i(t) + C_i q_i^2(t) = 0 \qquad (29)$$

$$i = 1,2,3,5,8.$$

For the intermediate plants, the situation is illustrated in Fig. 2. Let there be $(i - k)$ plants upstream from the $(i + 1)$st. The flow from each plant has a transport delay of τ_j $(j = k,\ldots,i)$ to the $(i + 1)$st plant. Then the $(i + 1)$st reservoir's dynamics are expressed by

$$\dot{S}_{i+1}(t) = i_{i+1}(t) + \sum_{j=k}^{i} q_j(t-\tau_j) - q_{i+1}(t). \tag{30}$$

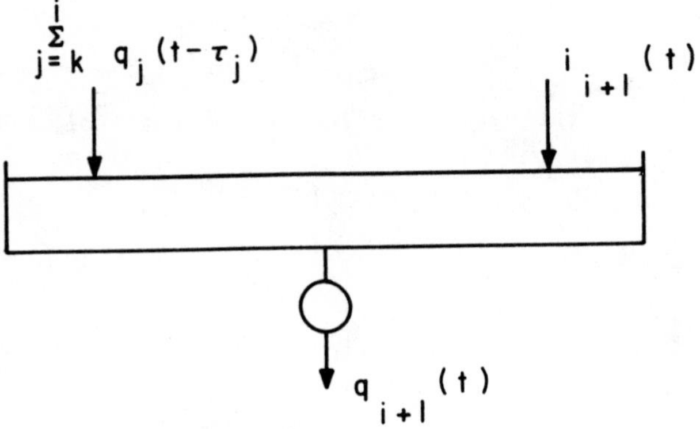

Fig. 2. The ith hydro plant's reservoir.

Integrating, one obtains

$$S_{i+1}(t) = D_{i+1}(t) + \sum_{j=k}^{i} x_j(t) - Q_{i+1}(t), \tag{31}$$

where

$$D_{i+1}(t) = S_{i+1}(0) + \int_0^t i_{i+1}(\sigma)\, d\sigma, \tag{32}$$

$$Q_i(t) = \int_0^t q_i(\sigma)\, d\sigma, \tag{33}$$

$$x_j(t) = \int_0^t q_j(\sigma - \tau_j)\, d\sigma, \qquad j = k,\ldots,i. \tag{34}$$

Let

$$\psi_j(t, \tau_j) = \int_{\tau_j}^{t-\tau_j} q_j(s)\, ds, \quad t \leq \tau_j, \quad j=k,\ldots,i, \tag{35}$$

which is a known function of time from the previous history of the system. Then (34) reduces to

$$x_j(t) = \begin{cases} \psi_j(t,\ \tau_j), & t \leq \tau_j,\quad i=k,\ldots,i, \qquad (36) \\ \psi_j(\tau_j,\ \tau_j) + Q_j(t-\tau_j), & t > \tau_j,\quad j=k,\ldots,i. \qquad (37) \end{cases}$$

The hydro power generated at the $(i + 1)$st plant is given by

$$P_{h_{i+1}}(t) = \frac{q_{i+1}(t)}{G_{i+1}} [\alpha_{i+1} + \beta_{y_{i+1}} S_{i+1}(t) - \beta_{T_{i+1}} q_{i+1}(t)]. \qquad (38)$$

Substituting (31) in (38) for $S_{i+1}(t)$, one obtains

$$P_{h_{i+1}}(t) + A_{i+1}(t)q_{i+1}(t) - B_{i+1}q_{i+1}(t) \sum_{j=k}^{i} x_j(t)$$

$$+ C_{i+1}q_{i+1}^2(t) + B_{i+1}q_{i+1}(t)Q_{i+1}(t) = 0. \qquad (39)$$

In the system at hand, plant number 4 is downstream from plants number 1, 2, and 3. Also plant number 6 is downstream from 4 and 5, and 7 is downstream from the 6th plant. Thus applying (39) we have

$$P_{h_4}(t) + A_4(t)q_4(t) - B_4q_4(t)[x_1(t) + x_2(t) + x_3(t)]$$

$$+ B_4q_4(t)Q_4(t) + C_4q_4^2(t) = 0, \qquad (40)$$

$$P_{h_6}(t) + A_6(t)q_6(t) - B_6q_6(t)[x_4(t) + x_5(t)]$$

$$+ B_6q_6(t)Q_6(t) + C_6q_6^2(t) = 0, \qquad (41)$$

$$P_{h_7}(t) + A_7(t)q_7(t) - B_7q_7(t)[x_6(t)] + C_7q_7^2(t)$$

$$+ \quad B_7q_7(t)Q_7(t) \quad = \quad 0. \tag{42}$$

C. Objective Functionals

Achieving the minimum total cost of supplying the power requirements of a system requires an accurate knowledge of the operating cost functions. A cost function represents the manner in which the total cost of operation of a generating unit varies with its output. The total cost of operation includes the fuel cost, cost of labor, supplies, and maintenance. No methods are presently available, however, for expressing the latter as a function of the output [1, 32]. Arbitrary methods of determining these costs are used. The most common one is to assume the cost of labor, supplies, and maintenance to be a fixed percentage of the incoming fuel costs. It is common practice to obtain the operating cost function by establishing the input-output curve for the plant considered, then adjusting for the cost of the fuel per unit input to the plant. All operating costs at the thermal plants are assumed to be given by

$$F_i(P_{s_i}(t)) \quad = \quad \alpha_i + \beta_i P_{s_i}(t) + \gamma_i P_{s_i}^2(t) \qquad \text{\$/hr.} \tag{43}$$

Here $P_{s_i}(t)$ is the active power generation at the ith thermal plant. The coefficients α_i, β_i, and γ_i are assumed available. Based on this, the objective of a classical economy dispatch problem for a system with m thermal plant is

$$\min_{P_{s_i}} (t), \qquad \int_0^{T_f} \sum_{i=1}^{m} F_i(P_{s_i}(t))\, dt. \tag{44}$$

Other possible objective functionals may be used. For example, in an all-hydro system, we may be interested in

$$\min_{P_{h_i}(t)} \int_0^{T_f} \sum_{i=1}^{N_h} A_i |P_{h_i}(t) - P^*_{h_i}|^2 \, dt. \tag{45}$$

Here $P^*_{h_i}$ denotes a desired active hydro power, say best efficiency point. Another alternative is to minimize the transmission losses $P_L(t)$, that is,

$$\min_{P_{s_i}, P_{h_i}} \int_0^{T_f} P_L(t) \, dt. \tag{46}$$

Another important objective in the power system's operation is its reliability. To improve the system reliability during operation, it is necessary to ensure that the reactive generations are minimally proportioned between the system generators [50, 51]. To achieve this it is suggested that the schedules obtained must

$$\min_{Q_{G_i}(t)} \int_0^{T_f} \left[\sum_{i=1}^{N_g} \sum_{j=1}^{N_g} Q_i(t) K_{ij} Q_j(t) \right] dt, \tag{47}$$

where the K_{ij}'s are assumed to be known weighting coefficients. The reactive power is denoted by Q_{G_i}. The duration of the optimization is T_f.

III. MULTIPLE CHAINS OF HYDRO PLANTS SYSTEM

A. The Problem

A hydrothermal electric power system is considered, which contains one thermal plant and eight hydro plants, as shown in

Fig. 1. The problem is that of short-range scheduling so a prediction of future system load and water supply is assumed available. The object here is to determine the individual active power generations in order to minimize the operating costs of the system under the following conditions:

(1) The operating costs are only due to the fuel cost at the thermal plant.

(2) The total active power generation in the system matches the load plus the transmission losses:

$$P_D(t) = P_{s_g}(t) + \sum_{i=1}^{8} P_{h_i}(t) - \sum_{i=1}^{9} \sum_{j=1}^{9} P_i(t) B_{ij} P_j(t) - \sum_{i=1}^{9} B_{i0} P_i(t). \tag{48}$$

(3) The volume of water discharge at any hydro plant over the optimization interval $[0, T_f]$ is a prespecified constant

$$\int_0^{T_f} q_i(\sigma)\, d\sigma = b_i, \qquad i=1,\ldots,8. \tag{49}$$

Here $q_i(t)$ is the rate of water discharge at the ith hydro plant.

(4) The hydro plants' active power generations are given by Eqs. (29) and (40)-(42).

B. A Minimum Norm Formulation

The objective functional to be minimized is

$$J_0 = \int_0^{T_f} [\alpha + \beta P_{s_g}(t) + \gamma P_{s_g}^2(t)]\, dt$$

subject to satisfying the constraints cited above. An augmented functional is obtained by including all the constraints but one in the cost functional using multiplier functions. The problem thus becomes that of minimizing

$$
\begin{aligned}
J(\cdot) = \int_0^{T_f} & [(\beta - \lambda(t)(1 - B_{g0}))P_{s_g}(t) \\
& + \sum_{i=1}^{8} [n_i(t) - \lambda(t)(1 - B_{i0})]P_{h_i}(t) \\
& + \sum_{i=1}^{6} [n_i(t)A_i(t) + m_i(t) + p_i(t, \tau_i)]q_i(t) \\
& + \sum_{i=7}^{8} [n_i(t)A_i(t) + m_i(t)]q_i(t) \\
& + \sum_{i=1}^{8} \dot{m}_i(t)Q_i(t) + \gamma P_{s_g}^2(t) \\
& + \lambda(t) \sum_{i=1}^{9} \sum_{j=1}^{9} P_i(t)B_{ij}P_j(t) \\
& + \sum_{i=1}^{8} C_i n_i(t)q_i^2(t) - \sum_{i=1}^{8} \frac{1}{2} B_i n_i(t)Q_i^2(t) \\
& - B_4 n_4(t)q_4(t)[x_1(t) + x_2(t) + x_3(t)] \\
& - B_6 n_6(t)q_6(t)[x_4(t) + x_5(t)] - B_7 n_7(t)q_7(t)x_6(t) \\
& + \sum_{i=1}^{6} r_i(t)x_i^2(t) + \sum_{i=1}^{6} - \theta_i(t, \tau_i)Q_i^2(t)]\, dt
\end{aligned}
\tag{50}
$$

subject to satisfying constraint (49).

The unknown function $\lambda(t)$ is the multiplier associated with the active power balance constraint. The hydropower generation constraints are included by using the functions $n_i(t)$. The relations

$$q_i(t) = \dot{Q}_i(t), \qquad i = 1,\ldots,8 \tag{51}$$

are included using the functions $m_i(t)$. The relations

$$x_j^2(t) = \begin{cases} \psi_j^2(t, \tau_j), & t \leq \tau_j \\ \psi_j^2(\tau_j, \tau_j) + 2\psi_j(\tau_j, \tau_j)Q_j(t-\tau_j) \\ \quad + Q_j^2(t-\tau_j), & t \geq \tau_j \end{cases} \tag{52}$$

are included using the functions $r_i(t)$. Note that Eqs. (52) are equivalent to Eqs. (36) and (37). The functions $\theta_i(t, \tau_i)$ and $p_i(t, \tau)$ are given by

$$\theta_i(t, \tau_i) = \begin{cases} r_i(t+\tau_i), & 0 \leq t \leq T_{\mathrm{f}} - \tau_i, \\ 0, & T_{\mathrm{f}} - \tau_i < t < T_{\mathrm{f}}, \end{cases} \tag{53}$$

$i = 1,\ldots,6,$

$$\dot{p}_i(t, \tau_i) = \begin{cases} 2\psi_i(\tau_i, \tau_i)r_i(t+\tau_i), & 0 \leq t \leq T_{\mathrm{f}} - \tau_i, \\ 0, & T_{\mathrm{f}} - \tau_i < t \leq T_{\mathrm{f}}, \end{cases} \tag{54}$$

$i = 1,\ldots,6.$

Define the control vector by

$$u(t) = \mathrm{col}[P(t), W_1(t), W_2(t), W_3(t), \ldots, W_8(t)] \tag{55}$$

with

$$P(t) = \mathrm{col}[P_{h_1}(t), \ldots, P_{h_8}(t), P_{s_g}(t)],$$

$$W_i(t) = \mathrm{col}[Q_i(t), q_i(t)], \qquad i = 1,2,3,5,8,$$

$$W_4(t) = \mathrm{col}[Q_4(t), q_4(t), x_1(t), x_2(t), x_3(t)],$$

$$W_6(t) = \mathrm{col}[Q_6(t), q_6(t), x_4(t), x_5(t)],$$

$$W_7(t) = \mathrm{col}[Q_7(t), q_7(t), x_6(t)].$$

We thus obtain the augmented objective functional Eq. (50) in vector form as

$$J(u) = \int_0^{T_f} L^T u(t) + u^T(t)B(t)u(t)\, dt. \tag{56}$$

Here the components of the vector $L(t)$ are the coefficients of the terms that are linear in the control components. The square symmetric matrix $B(t)$ is similarly the matrix of the coefficients of the terms quadratic in the control vector. Elements of $L(t)$ and $B(t)$ are functions of both the system parameters and the unknown multiplier functions.

The augmented functional is rewritten as

$$J(u) = \int_0^{T_f} [(u(t) + \tfrac{1}{2} V(t))^T B(t)(u(t) + \tfrac{1}{2} V(t)) - \tfrac{1}{2} V^T(t)B(t) \tfrac{1}{2} V(t)]\, dt \tag{57}$$

with

$$V^T(t) = L^T(t)B^{-1}(t). \tag{58}$$

The last term in the integrand of Eq. (57) does not depend explicitly on $u(t)$, so that it is only necessary to consider

$$J(u) = \int_0^{T_f} [(u(t) + \frac{1}{2} V(t))^T B(t)(u(t) + \frac{1}{2} V(t))]\, dt. \tag{59}$$

Define the 8×1 column vector as

$$b = \text{col}[b_1, \ldots, b_8] \tag{60}$$

and the 8×21 matrix K^T as

$$K^T = \begin{bmatrix} 0 & K_1^T & 0 & \\ 0 & 0 & K_2^T & \\ & & & \\ & & & K_8^T \end{bmatrix}$$

with

$$K_i^T = [0, 1], \quad i = 1,2,3,5,8, \qquad K_4^T = [0, 1, 0, 0, 0],$$

$$K_6^T = [0, 1, 0, 0], \qquad K_7^T = [0, 1, 0].$$

Then Eq. (49) reduces to

$$b = \int_0^{T_f} K^T u(s)\, ds. \tag{61}$$

The control vector $u(t)$ is considered an element of the Hilbert space $L_{2,B}^{(21)}[0, T_f]$ of the 21 vector-valued square integrable functions defined on $[0, T_f]$ endowed with the inner product definition

$$\langle V(t),\ u(t)\rangle = \int_0^{T_f} V^T(t)B(t)u(t)\ dt \tag{62}$$

for every $V(t)$ and $u(t)$ in $L_{2,B}^{(21)}[0,\ T_f]$, provided that $B(t)$ is positive definite.

The given vector b is considered an element of the real space R^8 with the Euclidean inner product definition

$$\langle X,\ Y\rangle = X^T Y$$

for every X and Y in R.

Equation (61) defines a bounded linear transformation $T\colon\ L_{2,B}^{(21)}[0,\ T_f] \to R^8$. This can be expressed as

$$b = T[u(t)], \tag{63}$$

and the cost functional given by Eq. (59) reduces to

$$J[u(t)] = \|u(t) + \frac{1}{2} V(t)\|^2. \tag{64}$$

Finally it is necessary only to minimize

$$J[u(t)] = \|u(t) + \frac{1}{2} V(t)\| \tag{65}$$

subject to

$$b = T[u(t)] \quad \text{for a given } b \text{ in } R. \tag{66}$$

C. The Optimal Solution

The optimal solution to the problem formulated using the results of Section I is

$$u_\xi(t) = T^\dagger[b + T(\frac{1}{2} V(t))] - \frac{1}{2} V(t) \tag{67}$$

where $T^{\dagger}$ is obtained as in the following

(1) T^*, the adjoint of T, is obtained using the identity

$$\langle \xi, Tu \rangle_{R^8} = \langle T^*\xi, u \rangle_{L_{2,B}^{(21)}[0, T_f]}. \tag{68}$$

(2) The operator J is evaluated from

$$J[\xi] = T[T^*\xi]. \tag{69}$$

(3) The inverse of J is next obtained.

(4) This yields the pseudoinverse operation given by

$$T^{\dagger}\xi = T^*[J^{-1}\xi]. \tag{70}$$

The details of the manipulations are omitted.

Componentwise the optimal solution is given by

$$P_{\xi}(t) = -\frac{1}{2} V_p(t), \tag{71}$$

$$Q_{\xi_i}(t) = \frac{\dot{m}_i(t)}{[B_i\dot{n}_i(t) + 2\theta_i(t, \tau_i)]}, \qquad i = 1,\ldots,6, \tag{72}$$

$$Q_{\xi_i}(t) = \frac{\dot{m}_i(t)}{B_i\dot{n}_i(t)}, \qquad i = 7,8. \tag{73}$$

This is, of course, in addition to the optimal expressions for the pseudocontrol variables $x(t)$ and $q(t)$ which will be eliminated together with the associated multipliers. The result is that the hydro variables will satisfy the general form

$$\frac{d}{dt}[2C_i n_i(t)\dot{Q}_{\xi_i}(t) + n_i(t)A_i(t)] + B_i n_i(t)Q_{\xi_i}(t)$$

$$+ \quad g_i(t) \quad = \quad 0. \tag{74}$$

Subject to the boundary conditions

$$Q_i(0) \quad = \quad 0, \qquad Q_i(T_f) \quad = \quad b_i. \tag{75}$$

Here the g_i's are given by

$$g_i(t) = B_4 n_4(t+\tau_i)\dot{Q}_{\xi_4}(t+\tau_i), \qquad 0 \le t \le T_f - \tau_i, \quad i=1,2,3,$$

$$g_i(t) = 0, \qquad T_f - \tau_i < t \le T_f, \quad i=1,2,3,$$

$$g_4(t) = B_6 n_6(t+\tau_4)\dot{Q}_{\xi_6}(t+\tau_4) - \frac{d}{dt}[B_4 n_4(t) \sum_{i=1}^{3} x_i(t, \tau_i)],$$

$$0 \le t \le T_f - \tau_4$$

$$= - \frac{d}{dt}[B_4 n_4(t) \sum_{i=1}^{3} x_i(t, \tau_1)],$$

$$T_f - \tau_4 < t \le T_f,$$

$$g_5(t) = B_6 n_6(t+\tau_5)\dot{Q}_{\xi_6}(t+\tau_5), \qquad 0 \le t \le T_f - \tau_6,$$

$$g_6(t) = 0, \qquad T_f - \tau_6 \le t \le T_f,$$

$$g_6(t) = B_7 n_7(t+\tau_6)\dot{Q}_{\xi_7}(t+\tau_6) - \frac{d}{dt}[B_6 n_6(t) \sum_{i=4}^{5} x_{\xi_i}(t, \tau_i)],$$

$$0 \le t \le T_f - \tau_6,$$

$$g_6(t) = -\frac{d}{dt}[B_6 n_6(t) \sum_{i=4}^{5} x_{\xi_i}(t, \tau_i)], \qquad T_f - \tau_6 < t \leq T_f,$$

$$g_7(t) = -\frac{d}{dt}[B_7 n_7(t) x_6(t)],$$

$$g_8(t) = 0, \tag{76}$$

We remark here that the introduction of pseudocontrol variables and justification of their use is given by Bauman [56].

D. Implementing the Optimal Solution

The optimal solution obtained so far involves the unknown multiplier functions $n_i(t)$ and $\lambda(t)$. These will be determined such that the active power balance and hydrogeneration constraints are satisfied. The resulting optimality equations are

$$P_{h_{i_\xi}}(t) = -V_{P_i}(t)/2, \qquad i=1,\ldots,8,$$

$$P_{s_{g_\xi}}(t) = -V_{P_g}(t)/2,$$

$$V_{P_i}(t) = 2[A_i(t)\dot{Q}_i(t) + B_i\dot{Q}_i(t)Q_i(t) + C_i\dot{Q}_i^2(t)], \quad i=1,2,3,5,8,$$

$$V_{P_4}(t) = 2[A_4(t)\dot{Q}_4(t) + B_4\dot{Q}_4(t)Q_4(t) + C_4\dot{Q}_4^2(t) - B_4\dot{Q}_4(t)[\sum_{i=1}^{3} x_i(t)]],$$

$$V_{P_6}(t) = 2[A_6(t)\dot{Q}_6(t) + B_6\dot{Q}_6(t)Q_6(t) + C_6\dot{Q}_6^2(t) - B_6\dot{Q}_6(t)[\sum_{i=4}^{5} x_i(t)]],$$

$$V_{P_7}(t) = 2[A_7(t)\dot{Q}_7(t) + B_7\dot{Q}_7(t)Q_7(t)$$
$$+ C_7\dot{Q}_7^2(t) - B_7Q_7(t)x_6(t)], \tag{78}$$

$$P_D(t) + 0.5\sum_{i=1}^{9} V_{P_i}(t) - 0.5\sum_{i=1}^{9} B_{i0}V_{P_i}(t)$$
$$+ 0.25\sum_{i=1}^{9}\sum_{j=1}^{9} V_{P_i}(t)B_{ij}V_{P_j}(t) = 0 \tag{79}$$

$$\ddot{Q}_{\xi_i}(t) + \rho_i(t)\dot{Q}_{\xi_i}(t) + \varepsilon_i\rho_i(t)Q_{\xi_i}(t) + G_i(\cdot) = 0,$$
$$i=1,\ldots,8, \tag{80}$$

where

$$\rho_i(t) = \dot{n}_i(t)/n_i(t), \qquad i=1,\ldots,8$$

$$\varepsilon_i(t) = B_i/2C_i, \qquad i=1,\ldots,8$$

$$G_i(\cdot) = g_i(\cdot) + [\dot{A}_i(t) + \rho_i(t)A_i(t)]/2C_i \qquad i=1,\ldots,8$$

with boundary conditions:

$$Q_{\xi_i}(0) = 0, \qquad Q_{\xi_i}(T_f) = b_i, \qquad i = 1,\ldots,8. \tag{81}$$

The discharge equations (80) and (81) are further rewritten as

$$\dot{Z}_i(t) = R_i(t)Z_i(t) + F_i(t), \qquad i = 1,\ldots,8, \tag{82}$$

$$MZ_i(0) + NZ_i(T_f) = C_i', \qquad i = 1,\ldots,8, \tag{83}$$

where

$$Z_i(t) = \text{col}[Q_{\xi_i}(t),\ \dot{Q}_{\xi_i}(t)], \tag{84}$$

$$R_i(t) = \begin{matrix} (0) & (1) \\ (-\varepsilon_i \rho_i(t)) & (-\rho_i(t)) \end{matrix},$$

$$F_i(t) = \text{col}[0, -G_i(\cdot)],$$

$$M = \begin{matrix} 1 & 0 \\ 0 & 0 \end{matrix}, \qquad N = \begin{matrix} 0 & 0 \\ 1 & 0 \end{matrix},$$

$$C_i' = \text{col}[0, b_i].$$

Equation (82) subject to the boundary conditions Eq. (83) is next transformed into the equivalent integral equation:

$$Q_{\xi_i}(t) = b_i t + \int_0^t s(T_\text{f} - t)f_i(s)\,ds + \int_t^{T_\text{f}} t[T_\text{f} - s]f_i(s)\,ds]/T_\text{f}, \quad i=1,\ldots,8, \tag{85}$$

$$\dot{Q}_{\xi_i}(t) = [b_i + \int_0^t - sf_i(s)\,ds + \int_t^{T_\text{f}} (T_\text{f} - s)f_i(s)\,ds]/T_\text{f}, \quad i=1,\ldots,8, \tag{86}$$

with

$$f_i(s) = \varepsilon_i \rho_i(s) Q_i(s) + \rho_i(s) Q_i(s) + G_i(s), \qquad i=1,\ldots,8. \tag{87}$$

Note that in Eqs. (85) and (86), the boundary conditions are

TABLE 1

HYDRO PLANTS' DATA

Plant	G ($ft^4\ hr^{-1}\ MW^{-1} \times 10^{-8}$)	$S\ Q$ ($ft^3 \times 10^{-12}$)	i ($ft^3\ hr^{-1} \times 10^{-6}$)	b ($ft^3 \times 10^{-9}$)	β_y ($ft^{-2} \times 10^9$)	C_T ($ft^{-2}\ hr \times 10^6$)
1	0.63	0.46	0.10	0.38	0.42	0.30
2	0.63	3.4	0.15	0.36	0.06	2.1
3	0.63	0.61	0.20	0.37	0.33	1.8
4	0.66	5.5	0.60	0.67	0.03	0.50
5	0.63	0.50	0.30	0.32	0.39	2.4
6	0.66	4.8	0.50	0.77	0.03	0.90
7	0.69	1.5	0.50	1.2	0.09	0.60
8	0.63	0.40	0.30	0.48	0.36	1.2

Plant hour[a]

Plant	1	2	3	4	5	6
1	0.16	0.15	0.15	0.30	0.16	0.30
2	0.32	0.30	0.30	0.60	0.32	0.60
3	0.48	0.45	0.45	0.90	0.48	0.90
4	0.64	0.60	0.60	1.20	0.54	1.20
5	0.80	0.75	0.75	1.50	0.80	1.50

[a] $\alpha_T = 0.0$. Initial condition function $\psi(t, \tau) \times 10^{-8}\ ft^3$.

Water transport delays are $\tau_1 = 2$ hr, $\tau_2 = 2$ hr, $\tau_3 = 3$ hr, $\tau_4 = 1$ hr, $\tau_5 = 1$ hr, and $\tau_6 = 2$ hr.

always satisfied.

The problem is thus reduced to that of solving Eqs. (78) and (79), and Eqs. (85) and (86). These equations can be written in the form

$$Z(t) = F[Z(t)], \tag{88}$$

where $Z(t)$ is a vector of the unknown functions. It is chosen here to use the modified contraction mapping defined by

$$Z_{k+1}(t) = [I - S]^{-1}[F(Z_k) - S(Z_k)], \tag{89}$$

where I is the identity operation, and S an operator such that $[I - S]$ is invertible. For a treatment of the procedures and convergence theorems we refer to the work of Falb and De Jong [57].

E. A Computational Example

A digital computer program was written to test the algorithm for an example system. The hydro plants' data are given in Table I. Tables II and III, give the transmission loss coefficients, thermal plant characteristics, and power demand, respectively.

Various values of the modified contraction mapping matrix S were tested. It appeared that $S = -I$ gives good convergence characteristics. The maximum relative error allowed was taken as $\varepsilon = 10^{-4}$. A plot of the maximum relative error (time-and-variable-wise) is shown in Fig. 3. The change in cost with iteration is given in Fig. 4. It was concluded that the program may be stopped after the eight iteration with a maximum relative error of 0.4×10^{-2}. With $\varepsilon = 10^{-4}$, the

TABLE II

TRANSMISSION LOSS AND FUEL COST COEFFICIENTS

$B_{11} = 0.20 \times 10^{-3}$	MW^{-1}	$B_{66} = 0.23 \times 10^{-3}$	MW^{-1}
$B_{22} = 0.19 \times 10^{-3}$	MW^{-1}	$B_{77} = 0.15 \times 10^{-3}$	MW^{-1}
$B_{33} = 0.21 \times 10^{-3}$	MW^{-1}	$B_{88} = 0.20 \times 10^{-3}$	MW^{-1}
$B_{44} = 0.20 \times 10^{-3}$	MW^{-1}	$B_{99} = 0.16 \times 10^{-3}$	MW^{-1}
$B_{55} = 0.15 \times 10^{-3}$	MW^{-1}	$B_{ij} = 0, \quad j \neq i$	
$\alpha = 0$	$\beta = 4.0$ \$/MW·hr	$\gamma = 0.0012$ \$/MW^2·hr	

TABLE III

ACTIVE POWER DEMAND

Hour	MW	Hour	MW	Hour	MW	Hour	MW
1	365	7	370	13	475	19	482
2	360	8	420	14	470	20	480
3	370	9	480	15	465	21	478
4	378	10	496	16	466	22	470
5	362	11	490	17	475	23	420
6	360	12	485	18	520	24	360

program takes 13 iterations to converge. Storage requirements are 96K. Plots of optimum power generation discharge and head are given in Figs. 5-7.

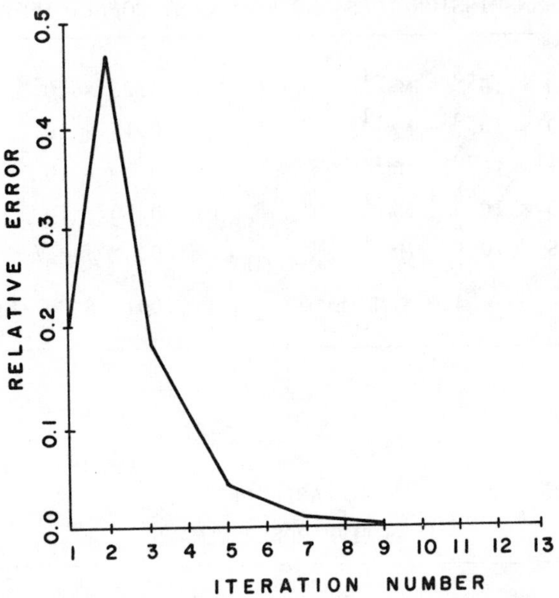

Fig. 3. Convergence characteristic for the test system.

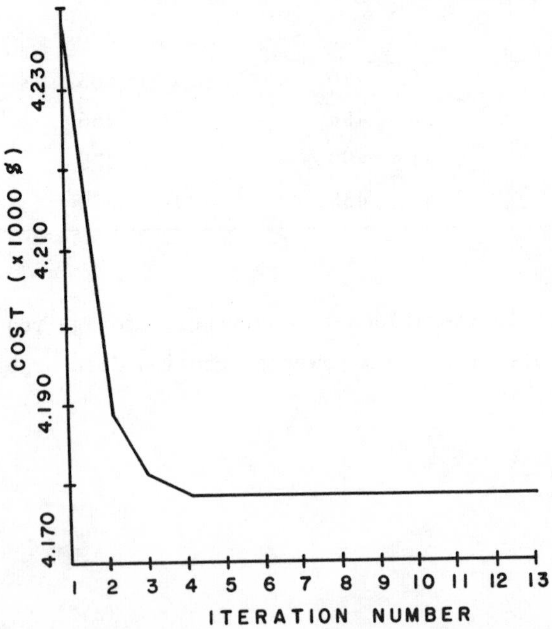

Fig. 4. Cost variation with iteration number.

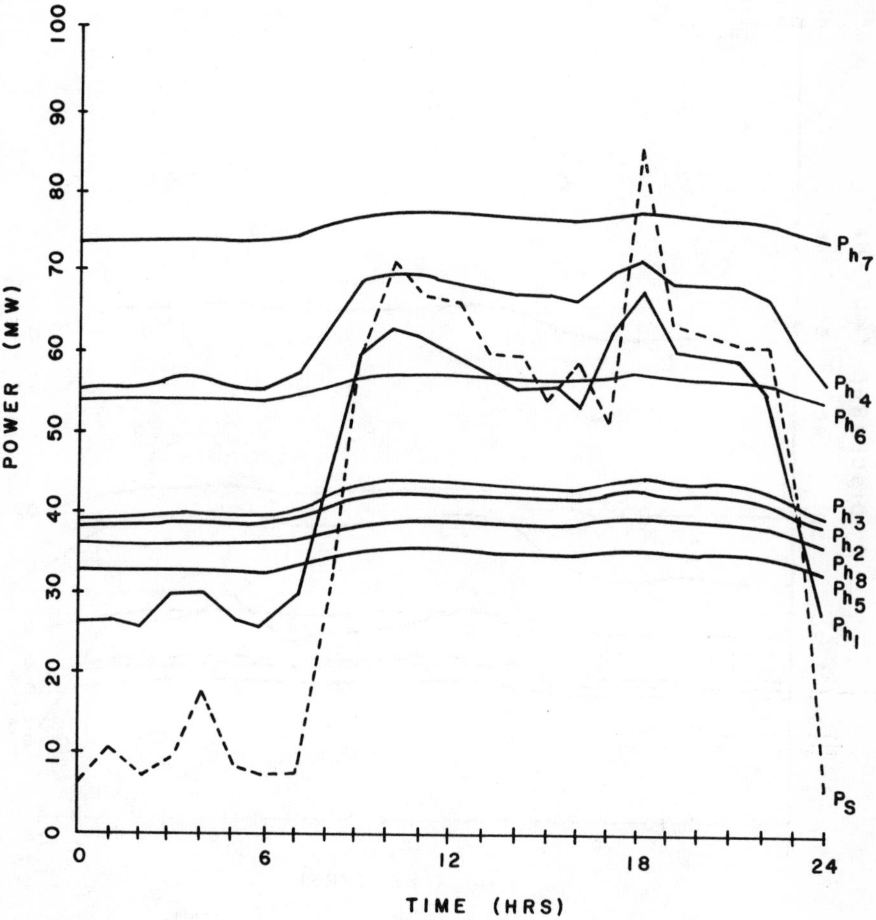

Fig. 5. Optimum power generations.

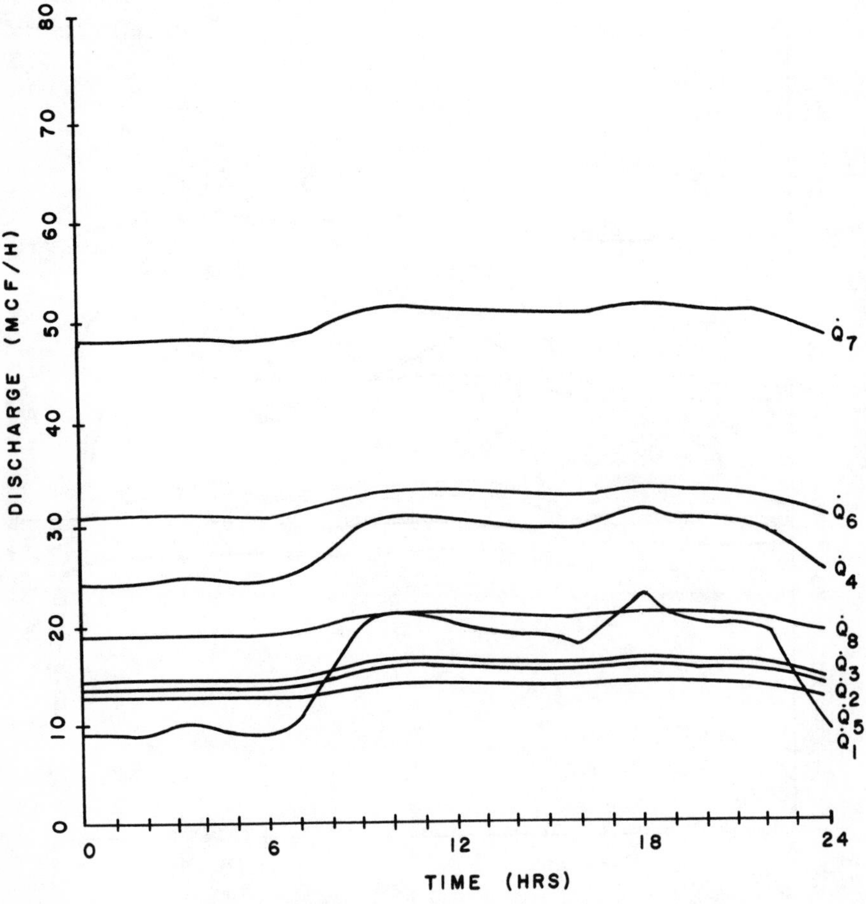

Fig. 6. Optimum rate of water discharges.

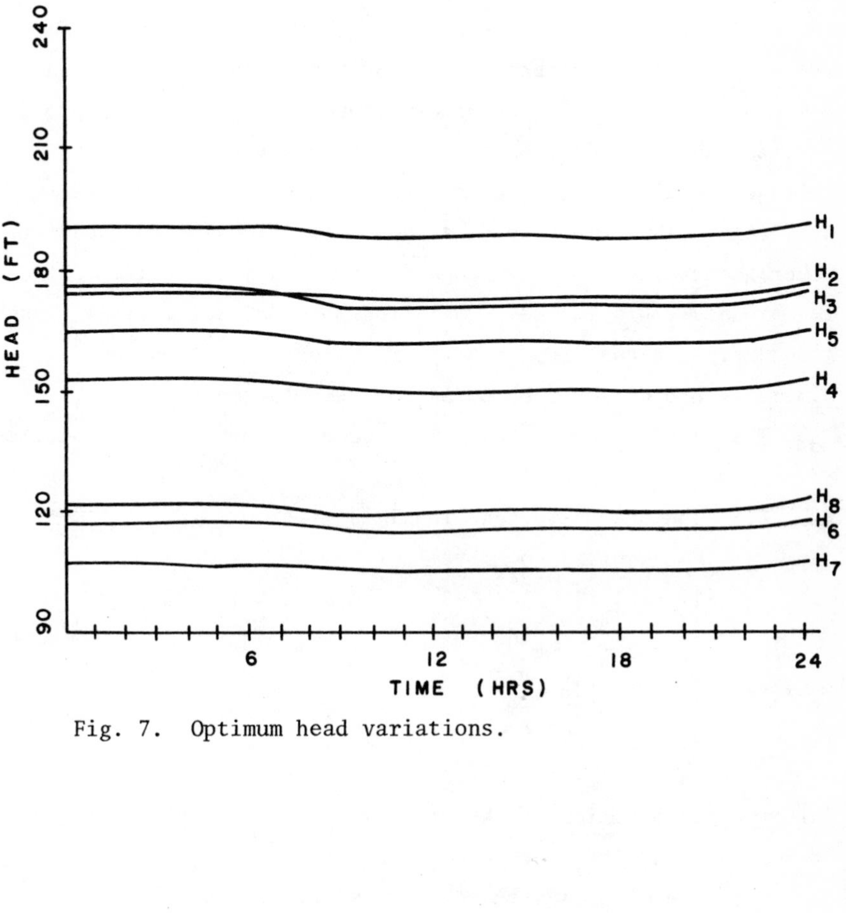

Fig. 7. Optimum head variations.

IV. OPTIMAL HYDROTHERMAL POWER FLOW

A. The Problem

A hydrothermal electric power system with an N-bus (node) network is considered. There are N_h hydro plants and $(N_g - N_h)$ thermal plants. Thus the system is assumed to have N_g generating plants (generator buses). The scheduling problem assumes that a prediction of active-reactive power demands, voltage profiles at generator buses, and water supply, are available. The object here is to determine active and reactive power generations in order to minimize a multiobjective cost functional. This includes operating costs at the thermal plants, in addition to ensuring that the reactive generations are minimally proportioned between the system's generators.

The problem formulated here assumes the following conditions:

(1) The objective functional is given by

$$J(\cdot) = \int_0^{T_f} \sum_{i=N_h+1}^{N_g} [\beta_i P_{s_i}(t) + \gamma_i P_{s_i}^2(t)] + \sum_{i=1}^{N_g} \sum_{j=1}^{N_g} Q_i(t) K_{ij} Q_j(t)\, dt. \tag{90}$$

This is a combination of the minimum operating cost functional of Eq. (44) and the functional of Eq. (47).

(2) The power system network is represented by the load flow Eqs. (8) and (9). In addition, the solutions sought will satisfy constraints given by Eq. (10) - (13).

(3) The hydro plants of the system are restricted (for simplicity of the presentation) to eight plants as shown in Fig. 1. Thus the active power generations of the hydro plants are given by Eqs. (29) and (40) - (42). Here we will replace

Q_i of these equations by Q_{w_i}. This is to avoid confusing the volume of water discharged with the reactive power.

(4) The volume of water discharge at any hydro plant over the optimization interval $[0, T_f]$ is a prespecified constant:

$$\int_0^{T_f} q_i(\sigma)\, d\sigma = b_i, \qquad i = 1,\ldots,N_h \tag{91}$$

for $N_h = 8$.

B. A Minimum Norm Formulation

An augmented objective functional is obtained by including the nonlinear constraints in the objective functional. This functional is given by

$$J_0(\cdot) = \int_0^{T_f} \sum_{i=1}^{10} I_{0_i}(\cdot), \tag{92}$$

$$\begin{aligned} I_{0_1}(\cdot) = [\sum_{i=1}^{N} \lambda_{P_i}(t)[-P_i(t) + E_{d_i}^2(t)G_i + E_{q_i}^2(t)G_i \\ - E_{d_i}(t) \sum_{\substack{j=1,\\ j\neq i}}^{N} [E_{d_j}(t)G^{ij} - E_{q_j}(t)B^{ij}] \\ - E_{q_i}(t) \sum_{\substack{j=1,\\ j\neq i}}^{N} [E_{d_j}(t)B^{ij} + E_{q_j}(t)G^{ij}]]], \end{aligned} \tag{93}$$

$$\begin{aligned} I_{0_2}(\cdot) = \sum_{i=1}^{N} \lambda_{q_i}(t)[Q_i(t) + E_{d_i}^2(t)B_i + E_{q_i}^2(t)B_i \\ + E_{q_i}(t) \sum_{\substack{j=1,\\ j\neq i}}^{N} [E_{d_j}(t)G^{ij} - E_{q_j}(t)B^{ij}] \end{aligned}$$

$$- E_{d_i}(t) \sum_{\substack{j=1,\\ j\neq i}}^{N} [E_{d_j}(t)B^{ij} + E_{q_j}(t)G^{ij}]], \tag{94}$$

$$I_{0_3}(\cdot) = \sum_{i=1}^{N_g-2} \lambda_{e_i}(t)[E_{d_i}^2(t) + E_{q_i}^2(t)], \tag{95}$$

$$I_{0_4}(\cdot) = \sum_{i=N_h+1}^{N_g} [\beta_i P_{s_i}(t) + \gamma_i P_{s_i}^2(t)]$$

$$+ \sum_{i=1}^{N_g} \sum_{j=1}^{N_g} Q_i(t)K_{ij}Q_j(t), \tag{96}$$

$$I_{0_5}(\cdot) = \sum_{i=1}^{N_g} M_i(t)[P_i^2(t) + Q_i(t) - S_i^{2^M}], \tag{97}$$

$$I_{0_6}(\cdot) = \sum_{i=1}^{N_g} \ell_i(t)[P_i^m - P_i(t)] \tag{98}$$

$$I_{0_7}(\cdot) = \sum_{i=1}^{N_g} \ell_i'(t)[P_i(t) - P_i^M], \tag{99}$$

$$I_{0_8}(\cdot) = \sum_{i=1}^{N_g} e_i(t)[Q_i^m - Q_i(t)], \tag{100}$$

$$I_{0_9}(\cdot) = \sum_{i=1}^{N_g} e_i'(t)[Q_i(t) - Q_i^M], \tag{101}$$

$$I_{0_{10}}(\cdot) = \sum_{i=1}^{8} n_i(t)P_{h_i}(t)$$

$$+ \sum_{i=1}^{6} [n_i(t)A_i(t) + m_i(t) + p_i(t, \tau_i)]q_i(t)$$

$$+ \sum_{i=7}^{8} [n_i(t)A_i(t) + m_i(t)]q_i(t)$$

$$+ \sum_{i=1}^{8} \dot{m}_i(t) Q_{w_i}(t)$$

$$+ \sum_{i=1}^{8} C_i n_i(t) q_i^2(t)$$

$$+ \sum_{i=1}^{8} - \frac{1}{2} B_i n_i(t) Q_{w_i}^2(t)$$

$$- B_4 n_4(t) q_4(t) [x_1(t) + x_2(t) + x_3(t)]$$

$$- B_6 n_6(t) q_6(t) [x_4(t) + x_5(t)]$$

$$- B_7 n_7(t) q_7(t) x_6(t)$$

$$+ \sum_{i=1}^{6} r_i(t) x_i^2(t) + \sum_{i=1}^{6} - \theta_i(t, \tau_i) Q_{w_i}^2(t)]. \tag{102}$$

We remark here that the interpretation of functions and pseudocontrols in $I_{0_{10}}(\cdot)$ is given in Section III. The inequality constraints are included using the Kuhn-Tucker theorem. The following exclusion equations must be satisfied at the optimum:

$$M_i(t)[P_i^2(t) + Q_i^2(t) - S_i^{2^M}] = 0, \tag{103}$$

$$\ell_i(t)[P_i^m - P_i(t)] = 0, \tag{104}$$

$$\ell_i'(t)[P_i(t) - P_i^M] = 0, \tag{105}$$

$$e_i(t)[Q_i^m - Q_i(t)] = 0, \tag{106}$$

$$e_i'(t)[Q_i(t) - Q_i^M] = 0, \tag{107}$$

for $i = 1,\ldots,N_g$, $t \in [0, T_f]$. Moreover, $\lambda_{P_i}(t)$, $\lambda_{q_i}(t)$, $\lambda_{e_i}(t)$, $n_i(t)$, and $m_i(t)$ are to be determined such that the corresponding equality constraints are satisfied.

We next eliminate terms explicitly independent of the control variables. The control variables here are $P_i(t)$ and $Q_i(t)$, $i = 1,\ldots,N_g$, $E_{d_i}(t)$, $i = 1,\ldots,N$, and $q_i(t)$, $Q_{w_i}(t)$, $i = 1,\ldots,N_h$, and $x_i(t)$. Further manipulations are carried out to recognize the fact that $E_{d_{N_g}}(t)$ and $E_{q_{N_g}}(t)$ are specified (this is the slack bus), and the assumption of zero phase angle at the slack bus. In the process, the following functions are defined:

$$a_{ij}(t) = -[\lambda_{p_i}(t)G^{ij} + \lambda_{q_i}(t)B^{ij}], \qquad i,j=1,\ldots,N, \quad i\neq j \tag{108}$$

$$b_{ij}(t) = -[\lambda_{p_i}(t)B^{ij} - \lambda_{q_i}(t)G^{ij}], \qquad i,j=1,\ldots,N, \quad i\neq j \tag{109}$$

$$a_{ii}(t) = \begin{cases} [\lambda_{p_i}(t)G_i + \lambda_{q_i}(t)B_i + \lambda_{e_i}(t)], & i=1,\ldots,N_g-2 \quad (110) \\ \lambda_{p_i}(t)G_i + \lambda_{q_i}(t)B_i, & i=N_g-1,\ldots,N. \quad (111) \end{cases}$$

Define the control vector as

$$u(t) = \text{col}[P(t), Q(t), E(t), W(t)] \tag{112}$$

with

$$P(t) = \text{col}[P_h(t), P_s(t)], \tag{113}$$

$$P_h(t) = \text{col}[P_1(t),\ldots,P_{N_h}(t)], \tag{114}$$

$$P_s(t) = \text{col}[P_{N_h+1}(t),\ldots,P_{N_g}(t)], \tag{115}$$

$$Q(t) = \text{col}[Q_1(t),\ldots,Q_{N_g}(t)], \tag{116}$$

$$E(t) = \text{col}[E_d(t),\ E_q(t)], \tag{117}$$

$$E_d(t) = \text{col}[E_{d_1}(t),\ \ldots,\ E_{d_{N_g-1}}(t),\ E_{d_{N_g+1}}(t),\ E_{d_N}(t)], \tag{118}$$

$$E_q(t) = \text{col}[E_{q_1}(t),\ldots,E_{q_{N_g-1}}(t),E_{q_{N_g+1}}(t),E_{q_N}(t)] \tag{119}$$

$$W(t) = \text{col}[W_1(t),\ldots,W_{N_h}(t)], \tag{120}$$

$$W_i(t) = \text{col}[Q_i(t),\ q_i(t)], \qquad i = 1,2,3,5,8, \tag{121}$$

$$W_4(t) = \text{col}[Q_4(t),\ q_4(t),\ x_1(t),\ x_2(t),\ x_3(t)], \tag{122}$$

$$W_6(t) = \text{col}[Q_6(t),\ q_6(t),\ x_4(t),\ x_5(t)], \tag{123}$$

$$W_7(t) = \text{col}[Q_7(t),\ q_7(t),\ x_6(t)]. \tag{124}$$

The control is a $2[n+N_h + N_g] \times 1$ column vector function. In our example system $N_h = 8$.

Define the auxiliary vector $L(t)$ as

$$L(t) = \text{col}[L_p(t),\ L_Q(t),\ L_E(t),\ L_W(t)], \tag{125}$$

$$L_p(t) = \text{col}[L_{p_h}(t),\ L_{p_s}(t)], \tag{126}$$

$$L_Q(t) = \text{col}[L_{Q_1}(t),\ldots,L_{Q_{N_g}}(t)], \tag{127}$$

$$L_{p_h}(t) = \text{col}[L_{p_1}(t),\ldots,L_{p_{N_h}}(t)], \tag{128}$$

$$L_{p_s}(t) = \text{col}[L_{p_{N_h+1}}(t),\ldots,L_{p_{N_q}}(t)], \tag{129}$$

$$L_E(t) = \text{col}[L_{E_d}(t),\ L_{E_q}(t)], \tag{130}$$

$$L_{E_d}(t) = \text{col}[L_{E_{d_1}}(t),\ldots,L_{E_{d_{N_g-1}}}(t),L_{E_{d_{N_g+1}}}(t),L_{E_{d_N}}(t)], \tag{131}$$

$$L_{E_q}(t) = \text{col}[L_{E_{q_1}}(t),\ldots,L_{E_{q_{N_g-1}}}(t),L_{E_{q_{N_g+1}}}(t)L_{E_{q_N}}(t)], \tag{132}$$

$$L_W(t) = \text{col}[L_{W_1}(t),\ldots,L_{W_{N_h}}(t)] \tag{133}$$

with

$$L_{p_i}(t) = [n_i(t) + \ell_i'(t) + \ell_i(t) - \lambda_{p_i}(t)], \quad i=1,\ldots,N_h, \tag{134}$$

$$L_{p_i}(t) = [\beta_i + \ell_i'(t) - \ell_i(t) - \lambda_{p_i}(t)], \quad i=N_h+1,\ldots,N_g, \tag{135}$$

$$L_{Q_i}(t) = [e_i'(t) - e_i(t) + \lambda_{q_i}(t)], \quad i=1,\ldots,N_g, \tag{136}$$

$$L_{E_{d_i}}(t) = [a_{N_g,i}(t) + a_{i,N_g}(t)]E_{d_{N_g}}(t), \quad i=1,\ldots,N,\ i\neq N_g \tag{137}$$

$$L_{E_{q_i}}(t) = -[b_{N_g,i}(t) + b_{i,N_g}(t)]E_{d_{N_g}}(t), \quad i=1,\ldots,N,\ i\neq N_g \tag{138}$$

$$L_{W_i}(t) = \text{col}[\dot{m}_i(t),\ m_i(t) + n_i(t)A_i(t) + p_i(t,\ \tau_i)], \quad i=1,2,3,5, \tag{139}$$

$$L_{W_4}(t) = \text{col}[\dot{m}_4(t),\ m_4(t) + n_4(t)A_4(t) + p_4(t,\ \tau_4),\ 0,\ 0,\ 0,], \tag{140}$$

$$L_{W_6}(t) = \mathrm{col}[\dot{m}_6(t),\ m_6(t) + n_6(t)A_6(t) + p_6(t,\ \tau_6),\ 0,\ 0,], \tag{141}$$

$$L_{W_7}(t) = \mathrm{col}[\dot{m}_7(t),\ m_7(t) + n_7(t)A_7(t),\ 0], \tag{142}$$

$$L_{W_8}(t) = \mathrm{col}[\dot{m}_8(t),\ m_8(t) + n_8(t)A_8(t)]. \tag{143}$$

Let the square matrix $B(t)$ be given by

$$B(t) = \mathrm{diag}[B_p(t),\ B_Q(t),\ B_E(t),\ B_W(t)] \tag{144}$$

with

$$B_p(t) = \mathrm{diag}[B_{p_{\mathrm{h}}}(t),\ B_{p_{\mathrm{s}}}(t)], \tag{145}$$

$$B_{p_{\mathrm{h}}}(t) = \mathrm{diag}[M_i(t)], \qquad i = 1,\ldots,N_{\mathrm{h}}, \tag{146}$$

$$B_{p_{\mathrm{s}}}(t) = \mathrm{diag}[B_{p_i}(t)], \qquad i = N_{\mathrm{h}}+1,\ldots,N_{\mathrm{g}}, \tag{147}$$

$$B_{p_i}(t) = M_i(t) + \gamma_i, \tag{148}$$

$$B_Q(t) = (K'_{ij}(t)) \tag{149}$$

$$K'_{ij} = K_{ij}, \qquad i \neq j, \quad i,j = 1,\ldots,N_{\mathrm{g}},$$

$$K'_{ii}(t) = M_i(t) + K_{ii}, \qquad i = 1,\ldots,N_{\mathrm{g}}, \tag{150}$$

$$B_W(t) = \mathrm{diag}[B_{W_i}(t)], \qquad i = 1,\ldots,N_{\mathrm{g}}, \tag{151}$$

$$B_{W_i}(t) = \mathrm{diag}[-\tfrac{1}{2} B_i n_i(t) + \theta_i(t,\ \tau_i),\ C_i n_i(t)],$$

$$i = 1,2,3,5, \tag{152}$$

$$B_{W_4}(t) = \operatorname{diag}\left[-\frac{1}{2} B_4 n_4(t) + \theta_4(t, \tau_4),\ B_{W_{qx_4}}(t)\right], \quad ($$

$$B_{W_{qx_4}}(t) = \begin{bmatrix} [C_4 n_4(t)] & [-B_4 n_4(t)/2] & [-B_4 n_4(t)/2] & [-B_4 n_4(t)/2] \\ [-B_4 n_4(t)/2] & [r_1(t)] & 0 & 0 \\ [-B_4 n_4(t)/2] & 0 & [r_2(t)] & 0 \\ [-B_4 n_4(t)/2] & 0 & 0 & [r_3(t)] \end{bmatrix}, \quad ($$

$$B_{W_6}(t) = \operatorname{diag}[-\frac{1}{2} B_6 n_6(t) + \theta_6(t, \tau_6),\ B_{W_{qx_6}}(t)], \quad ($$

$$B_{W_{qx_6}}(t) = \begin{bmatrix} (C_6 n_6(t)) & (B_6 n_6(t)/2) & (-B_6 n_6(t)/2) \\ (-B_6 n_6(t)/2 & r_4(t) & 0 \\ (-B_6 n_6(t)/2) & 0 & r_5(t) \end{bmatrix}, \quad ($$

$$B_{W_7}(t) = \operatorname{diag}[(-B_7 n_7(t)/2,\ B_{W_{qx_7}}(t)], \quad ($$

$$B_{W_{qx_7}}(t) = \begin{bmatrix} C_7 n_7(t) & -\frac{1}{2} B_7 n_7(t) \\ -\frac{1}{2} B_7 n_7(t) & r_6(t) \end{bmatrix}, \quad ($$

$$B_{W_8}(t) = \operatorname{diag}[(-\frac{1}{2} B_8 n_8(t),\ C_8 n_8(t)]. \quad ($$

Let

$$
B_{E_0}(t) = \begin{bmatrix}
a_{11}(t) & a_{12}(t) & & \cdots & a_{1N}(t) & 0 & -b_{12}(t) & -b_{13}(t) & \cdots & -b_{1N}(t) \\
a_{21}(t) & a_{22}(t) & & \cdots & a_{2N}(t) & -b_{21}(t) & 0 & -b_{23}(t) & \cdots & -b_{2N}(t) \\
\vdots & \vdots & & \vdots & \vdots & \vdots & \vdots & \vdots & \vdots & \vdots \\
a_{N1}(t) & a_{N2}(t) & & \cdots & a_{NN}(t) & -b_{N1}(t) & -b_{N2}(t) & -b_{N3}(t) & \cdots & 0 \\
0 & b_{12}(t) & b_{13}(t) & \cdots & b_{1N}(t) & a_{11}(t) & a_{12}(t) & & \cdots & a_{1N}(t) \\
b_{21}(t) & 0 & b_{23}(t) & \cdots & b_{2N}(t) & a_{21}(t) & a_{22}(t) & & \cdots & a_{2N}(t) \\
\vdots & \vdots & & \vdots & \vdots & \vdots & \vdots & & & \vdots \\
b_{N1}(t) & b_{N2}(t) & b_{N3}(t) & \cdots & 0 & a_{N1}(t) & a_{N2}(t) & & \cdots & a_{NN}(t)
\end{bmatrix}
$$

or

$$B_{E_0}(t) = (C_{ij}(t))_{2(N-1)\times 2(N-1)}, \qquad i,j \neq N_g, \quad i,j \neq N+N_g. \qquad (160)$$

Note that $B_{E_0}(t)$ is nonsymmetric. However, one can replace $B_{E_0}(t)$ by the symmetric matrix $B_E(t)$ such that

$$E^T(t)B_{E_0}(t)E(t) = E^T B_E(t)E(t),$$

where

$$B_E(t) = (b_{e_{ij}}(t))_{2(N-1)\times 2(N-1)}$$

with

$$b_{e_{ij}}(t) = b_{e_{ji}}(t),$$

$$b_{e_{ij}}(t) = \frac{1}{2}(C_{ij}(t) + C_{ji}(t)), \qquad i,j = 1,\ldots,2N,$$
$$j,j \neq N_g \text{ and } N+N_g.$$

Using these definitions, the cost functional of Eq. (92) reduces to

$$J_1[U] = \int_0^{T_f} L^T u(t) + u^T(t)B(t)u(t)\, dt. \qquad (161)$$

Let

$$V^T(t) = L^T(t)B^{-1}(t). \qquad (162)$$

Then Eq. (161) reduces to

$$J_1[U] = \int_0^{T_f} [(u(t) + \frac{1}{2} V(t))^T B(t)(u(t) + \frac{1}{2} V(t))$$
$$- \frac{1}{4} V^T(t)B(t)V(t)]\, dt. \qquad (163)$$

The last term in the integrand of Eq. (163) does not depend explicitly on $U(t)$, so it is only necessary to consider minimizing

$$J_2(\cdot) = \int_0^{T_f} [u(t) + \tfrac{1}{2} V(t)]^T B(t) [u(t) + \tfrac{1}{2} V(t)]\, dt \tag{164}$$

subject to satisfying Eq. (91) which is

$$b_i = \int_0^{T_f} q_i(t)\, dt, \qquad i = 1,\ldots,N_h. \tag{165}$$

Define the $N_h \times 1$ column vector

$$b = \mathrm{col}[b_1,\ldots,b_{N_h}] \tag{166}$$

and the matrix K^T such that Eq. (165) reduces to

$$b = \int_0^{T_f} K^T u(s)\, ds. \tag{167}$$

The dimension of the matrix K^T is compatible with those of b and u. Moreover all elements of this matrix are zero except for those corresponding to multiplication by the elements q_i of the control u, where the element is one.

The control vector $u(t)$ is considered to be an element of the Hilbert space $L_{2,B}^{N_T}[0, T_f]$ of the (N_T) vector-valued square integrable functions defined in $[0, T_f]$ endowed with the inner product definition

$$\langle V(t), u(t)\rangle = \int_0^{T_f} V^T(t) B(t) u(t)\, dt \tag{168}$$

for every $V(t)$ and $u(t)$ in $L_{2,B}^{N_T}[0, T_f]$, provided that $B(t)$ is positive definite.

The given vector b is considered an element of the real space R^{N_h} with the Euclidean inner product definition

$$<X, Y> = X^{T}Y$$

for every X and Y in R^{N_h}.

Equation (167) defines a bounded linear transformation $T\colon L_{2,B}^{N_T}[0, T_f] \to R^{N_h}$. This can be expressed as

$$b = T[u(t)]$$

and the cost functional given by Eq. (164) reduces to

$$J_2[u(t)] = \|u(t) + \frac{1}{2} V(t)\|^2.$$

Finally it is necessary only to minimize

$$J_2[u(t)] = \|U(t) + \frac{1}{2} V(t)\|$$

subject to $b = T[u(t)]$ for a given b in R^{N_h}. The dimension N_T of the control space is

$$N_T = 2[N_g + N - 1] + 22.$$

C. The Optimal Solution

The optimal solution to the problem formulated is given by

$$u_\xi = T^{\dagger}[b + T(\tfrac{1}{2}V(t))] - \frac{1}{2} V(t), \tag{169}$$

where $T^{\dagger}$ is obtained as outlined in Section III. C. The resulting optimal equations are

$$M_i(t)P_{h_{i_\xi}}(t) = -\frac{1}{2}[n_i(t) + \ell_i'(t) - \ell_i(t) - \lambda_{p_i}(t)],$$

$$i = 1,\ldots,N_\mathrm{h}, \qquad (170)$$

$$P_{S_{i_\xi}}(t) = -\frac{[\beta_i + \ell_i'(t) - \ell_i(t) - \lambda_{p_i}(t)]}{2[M_i(t) + \gamma_i]},$$

$$i = N_\mathrm{h}+1,\ldots,N_\mathrm{g}, \qquad (171)$$

$$Q_{i_\xi}(t) = -\frac{1}{2}\Big[\sum_{j=1}^{N_\mathrm{g}} [e_j'(t) - e_j(t) + \lambda_{q_j}(t)]d_{Q_{ij}}(t)\Big],$$

$$i = 1,\ldots,N_\mathrm{g}, \qquad (172)$$

$$E_{d_{i_\xi}}(t) = -\frac{1}{2}E_{d_{N_\mathrm{g}}}(t)\Big[\sum_{\substack{j=1,\\ j\neq N_\mathrm{g}}}^{N} (a_{N_\mathrm{g},j}(t) + a_{j,N_\mathrm{g}}(t))f_{ij}(t)$$

$$+ \sum_{\substack{j=1,\\ j\neq N_\mathrm{g}}}^{N} (b_{N_\mathrm{g},j}(t) + b_{j,N_\mathrm{g}}(t))h_{ij}(t)\Big], \qquad (173)$$

$$E_{q_{i_\xi}}(t) = -\frac{1}{2}E_{d_{N_\mathrm{g}}}(t)\Big[\sum_{\substack{j=1,\\ j\neq N_\mathrm{g}}}^{N} (a_{N_\mathrm{g},j}(t) + a_{j,N_\mathrm{g}}(t))h_{ij}(t)$$

$$- \sum_{\substack{j=1,\\ j\neq N_\mathrm{g}}}^{N} (b_{N_\mathrm{g},j}(t) + b_{j,N_\mathrm{g}}(t))f_{ij}(t)\Big], \qquad (174)$$

$$\frac{d}{dt}\,[2C_i n_i(t)\dot{Q}_{\xi_i}(t) + A_i(t)n_i(t)] + B_i n_i(t)Q_{\xi_i}(t) + g_i(t) = 0, \qquad i = 1,\ldots,8, \tag{175}$$

subject to the boundary conditions

$$Q_i(0) = 0, \qquad Q_i(T_f) = b_i.$$

The functions $g_i(t)$ are given by

$$g_i(t) = \begin{cases} B_4 n_4(t+\tau_i)\dot{Q}_{\xi_4}(t+\tau_i), & 0 \le t \le T_f - \tau_i \\ 0, & T_f - \tau_i \le t \le T_f \end{cases}$$

for $i = 1,2,3,$ and

$$g_4(t) = \begin{cases} B_6 n_6(t+\tau_4)\dot{Q}_{\xi_6}(t+\tau_4) - \frac{d}{dt}\,[B_4 n_4(t) \sum_{i=1}^{3} x_i(t,\, \tau_i)], & 0 \le t \le T_f - \tau_4 \\ -\frac{d}{dt}\,[B_4 n_4(t) \sum_{i=1}^{3} x_i(t,\, \tau_i)], & T_f - \tau_4 \le t \le T_f, \end{cases}$$

$$g_5(t) = \begin{cases} B_6 n_6(t+\tau_5)\dot{Q}_{\xi_6}(t+\tau_5) & 0 \le t \le T_f - \tau_6, \\ 0, & T_f - \tau_6 < t \le T_f, \end{cases}$$

$$g_6(t) = \begin{cases} B_6 n_6(t+\tau_6)\dot{Q}_{\xi_7}(t+\tau_6) - \dfrac{d}{dt}\left[B_6 n_6(t) \sum_{i=4}^{5} x_i(t, \tau_i)\right], & 0 < t < T_f - \tau_6, \\ -\dfrac{d}{dt}\left[B_6 n_6(t) \sum_{i=4}^{5} x_i(t, \tau_i)\right], & T_f - \tau_6 < t < T_f, \end{cases}$$

$$g_7(t) = -\frac{d}{dt}\left[B_7 n_7(t)\, x_6(t)\right],$$

$$g_8(t) = 0.$$

These equations together with the load flow equations, the exclusion equations, and the hydro power generation equations completely specify the optimal solution.

In manipulating the optimal equations, the following definitions were made:

(1) The matrix inverses $B_Q^{-1}(t)$ and $B_E^{-1}(t)$ are assumed to be

$$D_Q(t) = B_Q^{-1}(t), \qquad D_E(t) = B_E^{-1}(t),$$

with

$$D_Q(t) = (d_{Q_{ij}}(t)), \qquad D_E(t) = (d_{E_{ij}}(t)).$$

Note that the $d_{Q_{ij}}(t)$'s are functions of $M_i(t)$ and $d_{E_{ij}}(t)$'s are functions of $\lambda_{p_i}(t)$ and $\lambda_{q_i}(t)$.

(2) Recalling the symmetric matrix B_E, this can be written as

$$B_E(t) = \begin{bmatrix} A_E(t) & C_E(t) \\ C_E^T(t) & A_E(t) \end{bmatrix}$$

$$A_E(t) = (a_{ij_E}(t))_{(N-1)\times(N-1)},$$

$$C_E(t) = (C_{ij_E}(t))_{(N-1)\times(N-1)},$$

$$a_{ij_E}(t) = \frac{1}{2}(a_{ij}(t) + a_{ji}(t)),$$

$$C_{ij_E}(t) = \frac{1}{2}(-b_{ij}(t) + b_{ji}(t)).$$

Note that $A_E(t)$ is symmetric but $C_E(t)$ is not. Moreover $C_E(t) = -C_E^T(t)$. The inverse matrix $B_E^{-1}(t)$ is denoted by

$$D_E(t) = B_E^{-1}(t).$$

This turns out to be

$$D_E(t) = \begin{bmatrix} F_E(t) & H_E(t) \\ -H_E(t) & F_E(t) \end{bmatrix},$$

where

$$F_E(t) = [A_E(t) + C_E(t)A_E^{-1}(t)C_E(t)]^{-1}$$

$$H_E(t) = -A_E^{-1}(t)C_E(t)F_E(t).$$

V. TRAPEZOIDAL RESERVOIRS AND VARIABLE EFFICIENCY HYDRO PLANTS CONSIDERATIONS

A. The Problem

In formulating the problems in Sections III and IV, it was assumed that the efficiency of each hydro plant remains constant over the operating range. Another assumption is that of vertical-sided reservoirs at the hydro plants. In this section these two assumptions are relaxed and the modifications to both formulation and optimal solution are shown.

The ith hydro plant's active power generation is given by

$$P_{h_i}(t)G_i(t) = h_i(t)q_i(t). \tag{176}$$

This is precisely Eq. (16) except that here the inverse efficiency G_i is no longer a constant.

The effective hydraulic head at the ith hydro plant is given by

$$h_i(t) = y_i(t) - y_{T_i}(t). \tag{177}$$

The forebay elevation is related to the forebay volume of water stored $S_i(t)$ by

$$S_i(t) = \alpha_{y_i} y_i^2(t) + \beta_{y_i} y_i(t), \tag{178}$$

where α_{y_i} and β_{y_i} are constants for the trapezoidal reservoir given by

$$\alpha_{y_i} = \ell_i \tan \Phi_i, \qquad \beta_{y_i} = \ell_i b_{0_i}.$$

The geometry of a trapezoidal reservoir is shown in Fig. 8. The volume of water stored $S_i(t)$ is also given by

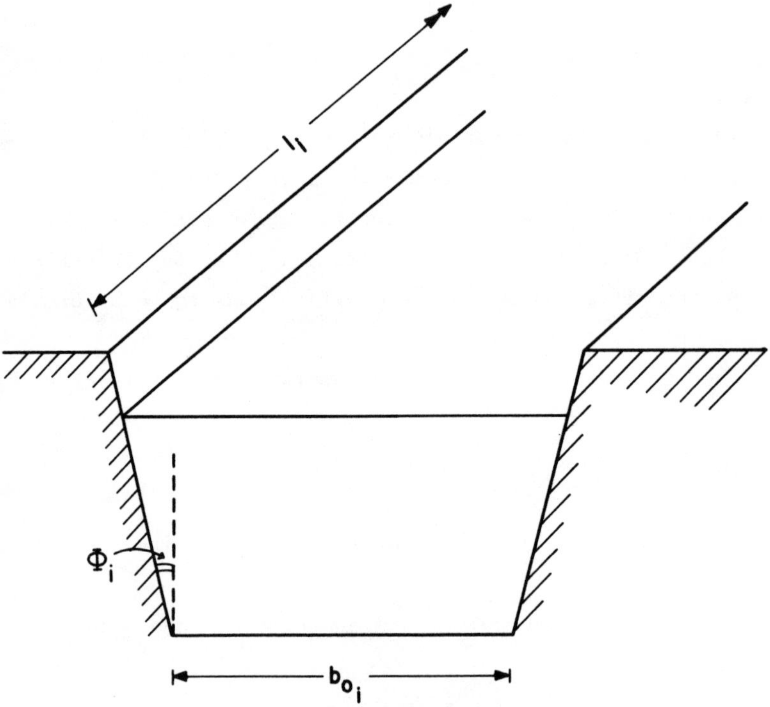

Fig. 8. A trapezoidal reservoir.

$$S_i(t) = S_i(0) + \int_0^t i_i(\sigma)\, d\sigma - \int_0^t q_i(\sigma)\, d\sigma. \tag{179}$$

This is the reservoir's dynamic equation in the case of no hydraulic coupling between the plants. Here $i_i(t)$ and $q_i(t)$ are the rates of water inflow and discharge, respectively. The volume of water stored variable $S_i(t)$ can be eliminated to obtain

$$\alpha_{y_i} y_i^2(t) + \beta_{y_i} y_i(t) + \int_0^t q_i(\sigma)\, d\sigma - S_i(0) - \int_0^t i_i(\sigma)\, d\sigma = 0. \tag{180}$$

The tail-race elevation $y_{T_i}(t)$ is given by

$$y_{T_i}(t) = y_{T_{i0}}(t) + \beta_{T_i} q_i(t). \tag{181}$$

Thus the hydro power generation equation becomes

$$P_{h_i}(t) G_i(t) + y_{T_{i0}} q_i(t) + \beta_{T_i} q_i^2(\ ) - q_i(t) y_i(t) = 0. \tag{182}$$

For all practical purposes the variation of the inverse efficiency $G_i(t)$ with the active power generation can be represented by

$$\alpha_{g_i} G_i^2(t) + \beta_{g_i} G_i(t) + \gamma_{g_i} P_{h_i}^2(t) + \delta_{g_i} P_{h_i}(t) + \theta_{g_i} = 0. \tag{183}$$

This is the equation of an ellipse and is assumed to hold true over the operating range $(P_i^m \leq P_{h_i}(t) \leq P_i^M)$ of the hydro plant. This is shown in Fig. 9. The performance of the ith hydro plant is completely specified by Eqs. (180), (182) and (183).

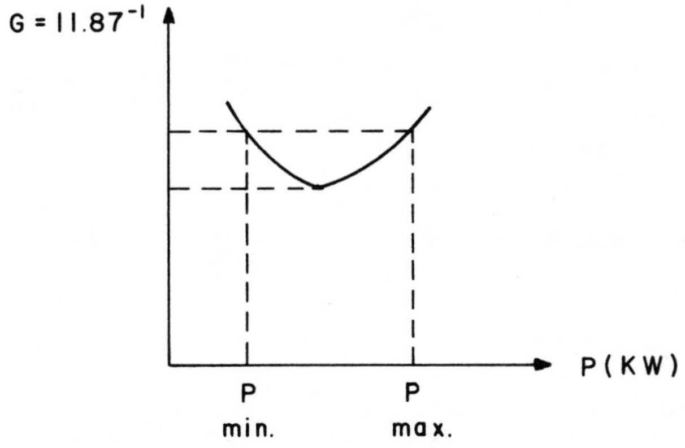

Fig. 9. Efficiency versus power output.

We will consider a hydrothermal electric power system. The electric network assumptions are precisely those of Section IV. In this treatment we will assume for brevity that the hydro plants of the system are on separate streams.

B. A Minimum Norm Formulation

With these assumptions, the only modification to the formulation of the problem in Section IV is using the three relations mentioned above. Note here that two more variables per plant are introduced; these are $y_i(t)$ and $G_i(t)$. Accordingly the functional $I_{0_{10}}(\cdot)$ in Eq. (102) becomes

$$\int_0^{T_f} I_{0_{10}}(t)\,dt = \int_0^{T_f} [\sum_{i-1}^{N_h} n_i(t)[P_{h_i}(t)G_i(t) + y_{T_{i0}}(t)q_i(t)$$

$$+ \beta_{T_i} q_i^2(t) - q_i(t)y_i(t)] + m_i(t)[\alpha_{g_i} G_i^2(t)$$

$$+ \beta_{g_i} G_i(t) + \gamma_{g_i} P_{h_i}^2(t) + \delta_{g_i}(t) + P_{h_i}(t) + \theta_{g_i}]$$

$$+ r_i(t)[\alpha_{y_i} y_i^2(t) + \beta_{y_i} y_i(t) + \int_0^t q_i(s)\,ds$$

$$- S_i(0) - \int_0^t i_i(\sigma)\,d\sigma]]\,dt. \tag{184}$$

Here the control variables are $P_{h_i}(t)$, $G_i(t)$, $q_i(t)$, and $y_i(t)$. Thus the terms explicitly independent of these variables can be dropped from $I_{0_{10}}(\cdot)$ so that one needs to consider only

$$I'_{0_{10}}(\cdot) = [\sum_{i=1}^{N_h} n_i(t)[P_{h_i}(t)G_i(t) + y_{T_{i0}}q_i(t)$$

$$+ \beta_{T_i}q_i^2(t) - q_i(t)y_i(t)] + m_i(t)[\alpha_{g_i}G_i^2(t)$$

$$+ \beta_{g_i}G_i(t) + \gamma_{g_i}P_{h_i}^2(t) + \delta_{g_i}P_{h_i}(t)]$$

$$+ \dot{r}_i(t)[\alpha_{y_i}y_i^2(t) + \beta_{y_i}y_i(t)] - r_i(t)q_i(t)]\, dt. \qquad (185)$$

The control vector defined in Eq. (112) is modified as

$$P_h(t) = \mathrm{col}[P_1(t), G_1(t), \ldots, P_{N_h}(t), G_{N_h}(t)], \qquad (186)$$

$$W_i(t) = \mathrm{col}[q_i(t), y_i(t)]. \qquad (187)$$

The last two equations correspond to Eq. (114) and Eq. (121), respectively. The control vector is a $2[n + 1.5N_h + N_g] \times 1$ column vector function in this case.

The auxiliary vector $L(t)$ of Eq. (125) is modified to

$$L_{P_h}(t) = \mathrm{col}[L_{P_1}(t), \ldots, L_{P_{N_h}}(t)], \qquad (188)$$

$$L_{P_i}(t) = \mathrm{col}[m_i(t)\delta_{g_i}(t) + \ell_i'(t) - \ell_i(t) - \lambda_{p_i}(t)),$$

$$(m_i(t)\beta_{g_i})], \qquad i = 1, \ldots, N_h \qquad (189)$$

$$L_{W_i}(t) = \mathrm{col}[(n_i(t)y_{T_{i0}} - r_i(t)), \dot{r}_i(t)\beta_{y_i}],$$

$$i = 1, \ldots, N_h. \qquad (190)$$

Finally the matrix $B(t)$ of Eq. (144) is modified to

$$B_{P_h}(t) = \operatorname{diag}(B_{P_i}(t)), \qquad i = 1,\ldots,N_h, \tag{191}$$

$$B_{P_i}(t) = \begin{bmatrix} [M_i(t) + m_i(t)\gamma_{g_i}] & \frac{1}{2}n_i(t) \\ \frac{1}{2}n_i(t) & m_i(t)\alpha_{g_i} \end{bmatrix}, \tag{192}$$

$$i = 1,\ldots,N_h,$$

$$B_{W_i}(t) = \begin{bmatrix} \beta_{T_i} n_i(t) & \frac{1}{2}n_i(t) \\ \frac{1}{2}n_i(t) & \alpha_{y_i}\dot{r}_i(t) \end{bmatrix}, \qquad i = 1,\ldots,N_h. \tag{193}$$

C. The Optimal Solution

The modification in deriving the optimal solution as obtained in Section IV follows easily. We will point out these modifications here.

The expression for $V(t)$ as given by Eq. (162) will not be changed. Only the expressions for its components $V_{P_h}(t)$ and $V_{W_i}(t)$ will be modified. Thus we have

$$V_{P_h}(t) = \operatorname{col}[V_{P_1}(t), V_{G_1}(t), \ldots, V_{P_{N_h}}, V_{G_{N_h}}(t)] \tag{194}$$

with

$$V_{p_i}(t) = ([m_i(t)\delta_{g_i} + \ell_i'(t) - \ell_i(t) - \lambda_{p_i}(t)]m_i(t)\alpha_{g_i}$$

$$- 0.5m_i(t)n_i(t)\beta_{g_i})/(\Delta_{W_{P_i}}(t)), \tag{195}$$

$$V_{G_i}(t) = ([M_i(t) + m_i(t)\gamma_{g_i}]m_i(t)\beta_{g_i} - 0.5n_i(t)$$

$$\times [m_i(t)\delta_{g_i} + \ell_i'(t) - \ell_i(t) - \lambda_{p_i}(t)])$$

$$\div \Delta_{W_{p_i}}(t), \tag{196}$$

$$\Delta_{W_{P_i}}(t) = m_i(t)\alpha_{g_i}[M_i(t) + \gamma_{g_i}m_i(t)] - \frac{1}{4}n_i^2(t).$$

Also

$$V_{W_i}(t) = \mathrm{col}[V_{W_{q_i}}(t), V_{W_{y_i}}(t)], \tag{197}$$

$$V_{W_{q_i}}(t) = [[n_i(t)y_{T_{i0}} - r_i(t)]\alpha_{y_i}r_i(t) + 0.5n_i(t)\dot{r}_i(t)$$

$$\times \beta_{y_i}]/\Delta_{W_i}(t), \tag{198}$$

$$V_{W_{y_i}}(t) = [0.5n_i(t)[n_i(t)y_{T_{i0}} - r_i(t)] + \beta_{T_i}\beta_{y_i}$$

$$\times n_i(t)r_i(t)]/\Delta_{W_i}(t). \tag{199}$$

The optimal solution as given by Eqs. (170)-(174) is unchanged. Only Eqs. (175) and (176) modify to

$$q_{\xi_i}(t) = [-\frac{1}{2} V_{W_{q_i}}(t)] + [b_i + \int_0^{T_f} [\frac{1}{2} V_{W_{q_i}}(t)]\, dt]\dot{r}_i(t)$$

$$\div [\Delta_{W_i}(t) \int_0^{T_f} \dot{r}_i(t)/\Delta_{W_i}(t)\, dt], \tag{200}$$

$$y_{\xi_i}(t) = [-\frac{1}{2} V_{W_{y_i}}(t)] + [[b_i + \int_0^{T_f} [\frac{1}{2} V_{W_{q_i}}(t)]\, dt]n_i(t)$$

$$\div [2\Delta_{W_i}(t) \int_0^{T_f} \dot{r}_i(t)/\Delta_{W_i}(t)\, dt]], \tag{201}$$

$$P_{h_{i_\xi}}(t) = -\frac{1}{2} V_{P_i}(t), \tag{202}$$

$$G_{i_\xi}(t) = -\frac{1}{2} V_{G_i}(t), \tag{203}$$

Here the variables $q_{\xi_i}(t)$, $y_{\xi_i}(t)$, $P_{h_i}(t)$ and $G_{i_\xi}(t)$ are to satisfy the equality constraints Eqs. (180), (182) and (183).

VI. CONCLUDING REMARKS

A. Conclusions

This chapter considered the applications of modern optimal control theory to problems of scheduling the operation of large-scale electric power systems. In particular a functional analytic optimization technique is applied to problems of economy scheduling of hydrothermal electric power systems. Here the minimum norm formulation is employed to find the optimum generation schedules. This investigation shows how the powerful minimum norm formulation can be applied to complex problems of high dimension. We remark here that the solution obtained is guaranteed to be the unique optimal solution. Moreover, limitations on the unknown functions obtained through this particular formulation facilitate the practical implementation of the optimal solution. A further simplification is the elimination of the multipliers associated with constraints that are linear in the control vector.

The problems presented here include many realistic situations. First, the time delay of flow between hydro plants on the same stream is included in the formulations. Also, the tail-race elevation effect on the operating hydraulic head is considered here. Second, the formulations in Sections IV and V are in terms of the exact model of the electric network. The reliability objective and the practical limitations on the network variables are also considered. Third, the effects of efficiency variations were incorporated and a trapezoidal reservoir is considered. These formulations in addition enjoy the advantages cited before.

An important aspect of the problems considered is the computational schemes adopted. Due to the nonlinearity of the resulting equations, the method is iterative in nature. Employing here the modified contraction mapping principles is very useful. Furthermore, the transformation of the differential equations into operator equations guarantees satisfaction of the boundary conditions at each iteration step.

The variety of power system optimum operation problems is evident from the example systems presented in this chapter. As the power systems engineer strives to achieve optimum operation, he is also concerned with the degree of model sophistication. The application of the principles presented in this chapter to other less sophisticated power system models is given by El-Hawary and Christensen [58-64].

B. Future Research

The minimum norm formulation employed in this investigation has demonstrated the capability of solving complex power system scheduling problems. Further research with the same technqiue would be desirable in order to explore the possibility of solving more complex problems. For example, it may be possible to solve common-flow problems in the case where the time delay of water flow is a function of the rate of water discharge. The justification of the constant time delay lies in the fact that these delays are of the order of a fraction of the optimization interval considered. Also efforts in the field of defining an overall reliability functional are highly desirable.

The question of optimum power system operation with respect to multiobjective functionals lacks an answer so far. It would be worth while to investigate these problems, taking stock from available theoretical results. Developing efficient simulators and on-line controllers for implementing the optimum strategies is another area of endeavour. It would also be interesting to develop a criterian indicating the relative merits and worth of various degrees of model sophistication.

REFERENCES

1. IEEE WORKING GROUP REPORT, IEEE Trans. Power Appar. Syst. PAS-90 (4), 1768-1775 (1971).
2. F. NOAKES and A. ARISMUNANDAR, AIEE Trans. Power Appar. Syst. PAS-81, 864-871 (1962).
3. C.W. WATCHORN, IEEE Trans. Power Appar. Syst. (1), 106-117 (1967).
4. E.B. DAHLIN, "Theoretical and Computational Aspects of Optimal Principles with Special Application to Power System Operation", Ph.D. Thesis, University of Pennsylvania, Philadelphia, (1964).

5. R.J. CYPSER, AIEE Trans., 73 (III-B) 1260-1267 (1954).

6. J.H. DRAKE, L.K. KIRCHMAYER, R.B. MAYALL, and H. WOOD, AIEE Trans., 81 (III-B) 242-250 (1962).

7. J. RICARD, Rev. Gen. Electric. 167 (1940).

8. W.G. CHANDLER, et al., AIEE Trans. 72 (III) 1057-1065 (1953)

9. A.F. GLIMN and L.K. KIRCHMAYER, AIEE Trans. 77 (III), 1070-1079 (1958).

10. J.J. CAREY, AIEE Trans. 73 (III-B) 1105-1111 (1954).

11. R.A. ARISMUNANDAR, "General Equations for Short Range Optimization of a Combined Hydrothermal Electric System", M.Sc. Thesis, University of British Columbia, Vancouver (1960).

12. R.A. ARISMUNANDAR and F. NOAKES, AIEE Trans., 82 (III) 88-93 (1962).

13. L.K. KIRCHMAYER and R.J. RINGLEE, IFAC Proc. Vol. 3 430/1-430/6 (1964).

14. B. BERNHOLTZ and L.V. GRAHAM, AIEE Trans., 80 (III), 921-932 (1960).

15. E.B. DAHLIN and D.W.C. SHEN, IEEE Trans. Power Appar. Syst. PAS-85 (5), 437-458 (1966).

16. I. HANO, Y. TAMURA, and S. NARITA, IEEE Trans. Power Appar. Syst. (5), 486-494 (1966).

17. H.A. BURR, "Load Division between Common-Flow Hydroelectric Stations", M.Sc. Thesis, M.I.T., Cambridge, Massachusetts, 1941.

18. P.R. MENON, "On Methods of Optimizing the Operation of Hydro-Electric Systems", Ph.D. Thesis, University of Washington, Seattle, 1962.

19. R.H. MILLER and R.P. THOMPSON, IEEE Power Eng. Soc., Winter Power Meeting, New York, Paper C72-5, 1972..

20. J. CARPENTIER, Bull. Soc. Franc. Elect., Ser. B, 8, 431-447 (1962).

21. J. PESCHON, D.S. PIERCY, W.F. TINNEY, O.J. TVEIT, and M. CUENOD, IEEE Trans. Power Appar. Syst. PAS-87, 40-48 (1968).

22. H.W. DOMMEL and W.F. TINNEY, IEEE Trans. Power Appar. Syst. PAS-87, 1866-1876 (1968).
23. G.W. STAGG and A.H. EL-ABIAD, "Computer Methods in Power System Analysis". McGraw-Hill, New York, 1968.
24. A.M. SASSON, IEEE Trans. Power Appar. Syst. PAS-88, 399-409 (1969).
25. A.H. EL-ABIAD and F.J. JAIMES, IEEE Trans. Power Appar. Syst. 413-422 (1969).
26. R.L. SULLIVAN and O.I. ELGERD, Power Industry Computer Appl. Conf., Denver, Colorado, (1969).
27. C.M. SHEN and M.A. LAUGHTON, Proc. IEE, 116, 225-239 (1969).
28. M. RAMAMOORTHY and J.G. RAO, Proc. IEE, 117, 794-798 (1970).
29. C.M. SHEN and M.A. LAUGHTON, Proc. IEE, 117, 2117-2127 (1970).
30. M.R. GENT and J.W. LAMONT, IEEE Trans. Power Appar. Syst. PAS-90, 2650-2660 (1971).
31. A.M. SASSON and H.M. MERRILL, IEEE Proc. 62, 9590972 (1974).
32. L.K. KIRCHMAYER, "Economic Operation of Power Systems". Wiley, New York, 1958.
33. F.M. KIRILLOVA, SIAM J. Control, 5, 25-50, (1967).
34. D.S. CARTER, Cana. J. Math. 9, 132-140 and 226, (1957).
35. W.T. REID, Duke Math. J. 29, 591-606 (1962).
36. A.V. BALAKRISHNAN, SIAM J. Control, 1 (2), 109-127 (1963).
37. L.W. NEUSTADT, SIAM J. Control, 1 (1), 16-31 (1962).
38. G.M. KRANC and P.E. SARACHICK, ASME Trans. J. Basic Eng., 85, 143-150 (1963).
39. W.A. PORTER, "A New Approach to the General Minimum Energy Problem", Joint Automat. Control Conf., Stanford, California, 1964.
40. W.A. PORTER and J.P. WILLIAMS, J. Math. Anal. Appl. 13, 251-264 (1966).
41. W.A. PORTER and J.P. WILLIAMS, J. Math. Anal. Appl. 13, 536-549 (1966).
42. W.A. PORTER, SIAM J. Control, 4 (3), 466-472 (1966).

43. W.A. PORTER, SIAM J. Control, 5 (4) 555-574 (1967).
44. W.A. PORTER, Math. System Theory, 5, (1), 20-44 (1971).
45. H.C. HSIEH, Adv. Control Systems, 2, 117 (1964).
46. W.A. PORTER, "Modern Foundations of Systems Engineering". MacMillan, New York, 1967.
47. A.N. KOLMOGROV and S.V. FOMIN, "Elements of the Theory of Functions and Functional Analysis, Vol. 1, Metric and Normed Spaces". Graylock, Albany, New York, 1957.
48. D.G. LUENBERGER, "Optimization by Vector Space Methods". Wiley, New York, 1969.
49. H.W. DOMMEL and W.F. TINNEY, IEEE Trans. Power Appar. Syst. PAS-87, 1866-1876 (1968).
50. O.I. ELGERD, "Electric Energy System Theory". McGraw-Hill, New York, 1971.
51. R.L. SULLIVAN, IEEE Trans. Power Appar. Syst., PAS-91, 906-910 (1972).
52. B. STOTT, IEEE Proc. 62, 916-929 (1974).
53. A.M. SASSON, "Economic Operation of Power Systems", Lecture notes presented at The Purdue Univ. Short Course in Power System Engineering, Lafayette, Indiana, 1971.
54. R.D. CHENOWETH, IEEE Power Energy. Soc., Winter Power Meeting, New York, Paper 71CP98PWR, 1971.
55. H.H. HAPP, J.F. HOHENSTEIN, L.K. KIRCHMAYER, and G.W. STAGG, IEEE Trans. Power Appar. Syst. PAS-83, 702-707 (1964).
56. E.J. BAUMAN, Adv. Control Systems, 6, 159 (1968).
57. P.L. FALB and J.L. De JONG, "Some Successive Approximation Methods in Control and Oscillation Theory". Academic Press, New York, 1969.
58. M.E. EL-HAWARY and G.S. CHRISTENSEN, IEEE Trans. Power Appar. Syst. PAS-91 (5), 1833-1839 (1972).
59. M.E. EL-HAWARY and G.S. CHRISTENSEN, IEEE Trans. Automatic Control, AC-17 (4), 518-521 (1972).
60. M.E. EL-HAWARY and G.S. CHRISTENSEN, Internat. J. Control, 16 (6), 1063-1072 (1972).

61. M.E. EL-HAWARY and G.S. CHRISTENSEN, Optimization Theory Appl. 12 (6), 576-587 (1973).

62. M.E. EL-HAWARY and G.S. CHRISTENSEN, IEEE Trans. Power Appar. Syst. PAS-92 (1), 356-364 (1973).

63. G.S. CHRISTENSEN and M.E. EL-HAWARY, IEEE Power Engrg. Soc., Summer Power Meeting, Vancouver, British Columbia, Paper C73450-5, 1973.

64. M.E. EL-HAWARY and G.S. CHRISTENSEN, IEEE Conf. Expo., Toronto, Paper No. 75012, 8-9, 1975.

A NEW APPROACH TO HIGH-SPEED TRACKED VEHICLE SUSPENSION SYNTHESIS

CHRISTIAN GUENTHER

Messerschmitt-Bolkow-Blohm
Munich, West Germany

I. INTRODUCTION

Future tracked ground transportation systems will require vehicles with a maximum speed of up to 500 km/hr. A two-stage suspension system consisting of a primary suspension that supports and guides the vehicle truck and a secondary suspension supporting the vehicle body is featured by almost all high-speed vehicle configurations.

While much attention has been paid to intricate primary suspension systems such as magnetic levitation or advanced steel wheel-steel rail suspension it is also necessary to develop new concepts for the secondary suspension because the requirements on its performance are higher as the vehicle design speed increases.

In the following a new approach to the derivation of the structure of a secondary suspension will be shown. After developing the general suspension configuration the associated deterministic and stochastic control problems will be stated and solved.

II. TRACKED HIGH-SPEED VEHICLE SECONDARY SUSPENSION REQUIREMENTS

The requirements of a secondary suspension system can be stated as follows:

(1) The suspension must isolate passengers from disturbances caused by guideway irregularities, structural vibrations and aerodynamic forces such that given ride quality criteria are fulfilled. Such criteria are formulated for stationary and nonstationary conditions. The most common criterion for the former is based on the impressions of persons subjected to sinusoidal vibrations, but a more accurate one should include the interpretation of the guideway irregularity as a sample function of a stochastic process. The criteria for centrifugal force unbalance and other nonstationary conditions can be defined as maximum allowable transient acceleration limits.

(2) The suspension must maintain a desired vehicle body attitude under influence of loading disturbances, aerodynamic forces, propulsion forces, and acclerations caused by the guideway curvature. This requirement is often violated if a suspension is optimized only with respect to Requirement (1).

(3) The suspension system must assure that the vehicle follows the guideway in a desired fashion. This requirement can be decomposed into two parts: First, the vehicle must follow the guideway in stable manner; and second, the suspension must tilt the vehicle body if the vehicle traverses curves not at the so-called equilibrium speed.

(4) The suspension must operate under given vehicle component displacement constraints.

(5) The dynamic and static suspension characteristics must be invariant under vehicle body mass and inertia variations. This requirement assures that demands (1)-(4) are fulfilled regardless

whether the vehicle is fully or only partially loaded. The optimal suspension must, therefore, have adaptive features.

III. VEHICLE SECONDARY SUSPENSION SYNTHESIS PROBLEM STATEMENT

In order to concentrate on the essentialities of the new approach we will adopt a problem statement that contains assumptions that in an actual suspension design process would have to be checked. We will state the vehicle secondary suspension synthesis problem statement.

Given is a symmetric rigid vehicle consisting of a body and two trucks having a primary suspension that can be approximated by a spring with constant spring rate and a controllable preload with no unsprung mass (Maglev vehicle) traveling on a rigid guideway consisting of vertical and lateral curves and ramps. Assuming that the vehicle speed is constant, sensor and actuator dynamics can be neglected, no control power limitation is given, the secondary suspension performance is not influenced by truck yaw and pitch motions, and external forces are measurable, we will synthesize a secondary suspension system that fulfills the suspension requirements stated in Section II. Assume further that the vehicle configuration is such that vehicle body tilting and guideway tracking are not limited by displacement constraints. However, the response time of the tilting system must be smaller or equal to a given maximum response time T_{rmax} and the step response must exhibit a specified damping behavior.

The vehicle body acceleration minimization function has to be performed within given displacement constraints of the relative motion between the vehicle body and the trucks, and between the trucks and the guideway, for a guideway irregularity spectral density specified in lateral and vertical direction. The displacement limits are given under consideration of the requirements for vehicle body tilting. They can, therefore, be fully utilized for vibration isolation only.

The resulting synthesized suspension should be time invariant except for the part that maintains nominal mass and inertia parameters of the vehicle body and must not require preview sensors. Assume that only instantaneous feedback is available; the states are measurable only in certain linear combinations. No state estimation should be used to determine unmeasurable states. However, sensors that can measure the major external forces are admissible.

IV. DECOMPOSITION OF THE VSSSPS

The five suspension requirements of the VSSSPS will now be regrouped in order to perform a decomposition of the VSSSPS which will result into the general structure of the secondary suspension to be synthesized.

Consider Requirement (3), the guideway tracking requirement. Whereas the truck must follow the actual guideway surface and must maintain the primary air gaps or contact forces within given limits the interpretation of the vehicle body tracking requirement adopted here is such that the expected value of the vehicle body position must follow the mean order "ideal" guideway geometry which is followed by the moving track frame T. Tilting of the vehicle body in curves to cancel the lateral component of the resulting acceleration vector must also be understood as controlling the mean of the vehicle body roll angle and not its actual instantaneous value.

In order to fulfill Requirement (1) the deviations of certain vehicle body states with respect to their mean must be minimized, that is, the covariance of certain states must be minimized in order to achieve the desired ride quality.

Finally consider Requirement (2), the external force cancellation requirement. If all external forces are measurable, then no vehicle body state would be required for the open-loop cancellation of these forces. Indeed, vehicle body and passenger weight, propulsion and braking forces are easily measurable. The track geometry acceleration can be shown to be measurable since the mean of the inertial truck acceleration and the primary suspension gab are both measurable. We shall assume that the aerodynamic force acting on the vehicle can be measured using an air data computer and sensors distributed over the vehicle surface, placed and calibrated by wind tunnel tests. It is, therefore, feasible to assume for the synthesis phase of the suspension design that all major external forces are measurable.

Using the above interpretations we can regroup Requirements (1)-(5) into three disjoint sets I, II, and III as shown in Table I. None of the sets contains incompatible requirements. Requirement Set I can be fulfilled with an open-loop control using the measurable external forces, the translational and rotational vehicle body acceleration vectors (preload-and-mass control). The set of admissible controls to fulfill Requirement Set II is the set containing means $E(x)$ of the vehicle state x (tracking control), a subset of the zero-mean stochastic state variable $\hat{x} \triangleq x - E(x)$ (vibration control). Note that x is defined with respect to the moving track frame T.

The regrouping of the vehicle suspension requirements leads, therefore, to a decomposition of the VSSSPS as shown in Fig. 1. It is seen that the decomposition is performed at two levels that results in three independent subproblems in the sense that three independent sets of control are used in order to fulfill Requirements I, II and III, respectively. The incompatibility of vibration isolation (demanding a "soft" suspension) and tracking (demanding a "stiff" suspension) has been removed by this structure.

V. GENERAL STRUCTURE OF THE SYNTHESIZED SUSPENSION

Figure 2 illustrates the general structure of the synthesized suspension using the proposed decomposition approach. The three parts of the suspension and their generation are schematically

TABLE I

REGROUPING OF VEHICLE SUSPENSION REQUIREMENT

Original Vehicle-Suspension Requirements		New Requirement Sets		
Requirement Number	Purpose	I	II	III
1	Vibration isolation for vehicle body			×
2	External force effect cancellation	×		
3	Vehicle body tilting and guideway tracking		×	
	Truck tracking of guideway			×
4	Displacement constraints			×
5	Maintenance of nominal mass and inertia of vehicle body	×		

shown. The preload-and-mass control u_{pr} is a nonlinear function of the external forces, vehicle acceleration, and vehicle body mass parameters. The tracking control $\bar{u}^0$ is a linear combination of the error between the command vector c and the mean value of the accessible vehicle states. Finally, the vibration control $\hat{u}^0$ is seen to be a linear combination of the difference between the actual instantaneous values of the accessible states, relative accelerations (vehicle with respect to guideway), and their means.

If the optimal control methods would have been applied without any decomposition, that is, the admissible controls are functions of the state vector only, the following facts effectively

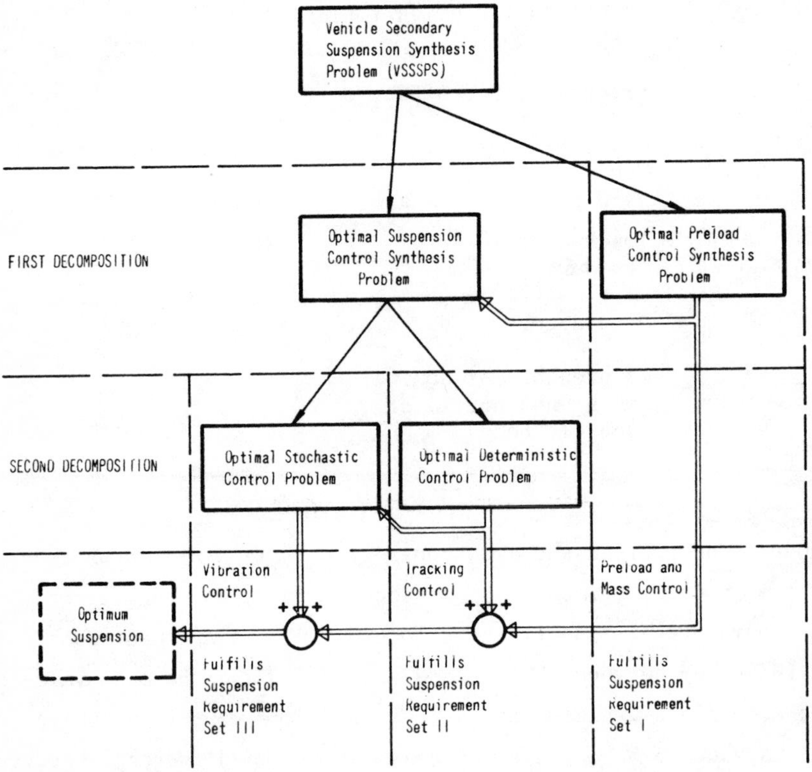

Fig. 1. Proposed decomposition of the high-speed tracked vehicle secondary suspension synthesis problem (VSSSPS) and relation of the subproblems to the vehicle suspension requirements.

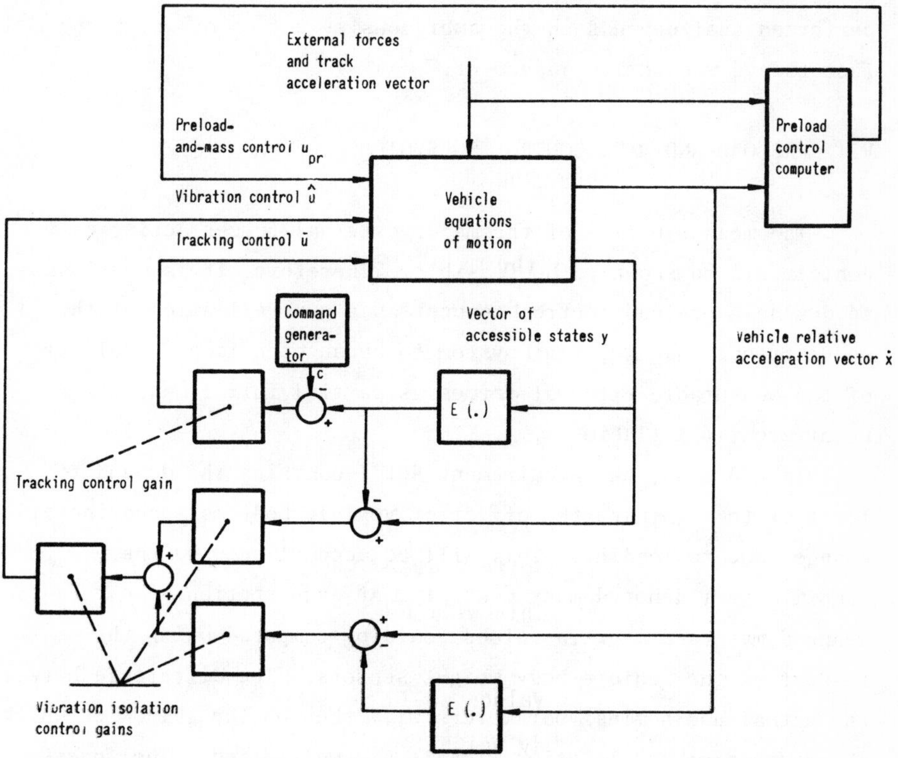

Fig. 2. General structure of the synthesized secondary suspension.

prevent an optimal solution:

(1) The plant equation is nonlinear and time-varying, thus preventing the solution with reasonable amount of computer time.

(2) A cost functional would have to be generated such that external force cancellation, vibration isolation, and guideway tracking is assured by a single control vector operating on the state vector x. This means that because of the incompatibility of the requirements a compromise optimization would have to be performed that depends on the ambiguous selection of weighting factors and subjective judgement.

VI. PRELOAD-AND-MASS CONTROL SUBSYSTEMS

The measurability of the major external forces acting on the vehicle can be shown to be feasible. Therefore, it is possible to define a preload control concept as a generalization of the conventional load-leveling system to be an open-loop cancellation of the measurable external forces as partial fulfillment of Requirement (1) (Table I).

In addition, the Requirement Set I contains the demand to limit or to eliminate the effect of vehicle body mass and inertia changes due to loading. This will be accomplished by the adaptive part denoted mass control. At each station stop the changed mass and inertia values could be computed using the readout of the vehicle body weight sensors. The difference between the actual and nominal parameters will then be the gain with which the respective acceleration signals are multiplied. The forces generated by the mass control part of the preload control system must be realizable; that is, the resulting primary suspension forces must be achievable within the displacement constraints given in the vehicle suspension problem statement and considering the effect of Earth's gravity. It turns out that the acceleration signals used for the mass control, the relative acceleration of vehicle body to guideway, are typically less than 1/10 of the Earth's gravity since the vibration isolation part of the suspension tends to minimize these accelerations. Therefore, it can be concluded that the mass control is feasible.

Because of the concept's open-loop characteristic it is not necessary to solve the vehicle equations of motion to calculate the optimal preload-and-mass control forces. The vehicle stability is not affected since besides the external forces only vehicle body acceleration signals are used in the secondary suspension. And unbalanced forces will be reacted by the tracking-control and vibration-isolation subsystems. Hence an error in force cancellation will cause a rather small squat because of the characteristics of these subsystems. In other words the preload-and-mass control reduces the effect of external forces and mass variations such that neglecting the residual effects is justified for the synthesis of the tracking-control and vibration-isolation subsystems.

VII. TRACKING CONTROL AND VIBRATION ISOLATION SUBSYSTEM

A. Decomposition of the Vehicle Plant Equation

The secondary suspension control forces vectors F_i, $i = 1, \ldots, 4$, and the secondary suspension roll torque vectors L_1 and L_2 by which the two trucks are connected to the vehicle body can be decomposed into two parts using the definitions

$$\hat{F}_i \triangleq F_i - P_i^0, \quad i = 1,\ldots,4; \tag{1}$$

$$\hat{L}_i \triangleq L_i - T_i^0, \quad i = 1,2, \tag{2}$$

where P_i^0 and T_i^0 are the secondary preload force and preload torque vectors, respectively. The suspension force vectors $\hat{F}_i$

and torque vectors $\hat{L}_i$ will be determined by the optimal control problems to be defined.

The primary suspension force vectors F_{pi}, $i = 1,\ldots,4$, and the primary suspension roll torque vectors L_{p1} and L_{p2} can be similarly decomposed as

$$\hat{F}_{pi} \triangleq F_{pi} - P^0_{pi}, \qquad i = 1,\ldots,4; \tag{3}$$

$$\hat{L}_{pi} \triangleq L_{pi} - T^0_{pi}, \qquad i = 1,2, \tag{4}$$

where P^0_{pi} and T^0_{pi} are the primary preload force vectors and preload roll torque vectors, respectively. Both of them are controllable as defined in Section III.

Based on the underlying vehicle model the primary and secondary preload vectors can be calculated without solving the equations of motion. Substituting them back into the vehicle guideway equations of motion results in a set of linear equations which will have the form

$$\dot{x} \triangleq \begin{bmatrix} \mathring{x}_V \\ \mathring{x}_T \end{bmatrix} = \begin{bmatrix} A_V & \bar{0} \\ \bar{0} & A_T \end{bmatrix} \begin{bmatrix} x_V \\ x_T \end{bmatrix} + \begin{bmatrix} B_V \\ B_T \end{bmatrix} u + \begin{bmatrix} \bar{0} \\ -B_T \end{bmatrix} (f_p + u_p), \tag{5}$$

Once x_V and x_T are defined then $\mathring{x}_V$ and $\mathring{x}_T$ are known in all literature as their derivative in which x_V denotes the vehicle body state vector, x_T the truck state vector, u the secondary suspension force vector consisting of the nonzero elements of $\hat{F}_i$ and $\hat{L}_i$, f_p the primary suspension force vector consisting of the nonzero elements of $\hat{F}_{pi}$ and $\hat{L}_{pi}$, and finally u_p the preload compensation vector consisting of the nonzero elements of the vectors U_{Pi} and U_{Ti}. The latter have been added to the primary preload force vector P^0_{pi} and preload roll torque vector T^0_{pi} in order to react the tracking control forces acting on the trucks. Finally, A_V, B_V, A_T, and B_T are constant matrices of appropriate dimensions. Note that from now on differentiation with respect to the moving track frame T will be denoted as defined in Eq. (5).

The elements of f_p are nothing else than the primary suspension spring forces, hence Eq. (5) can be rewritten to arrive at the final form of the vibration and tracking control plant equation of the vehicle:

$$\dot{x} \triangleq \begin{bmatrix} \mathring{x}_V \\ \mathring{x}_T \end{bmatrix} = \begin{bmatrix} A_V & \bar{0} \\ \bar{0} & A_p \end{bmatrix} \begin{bmatrix} x_V \\ x_T \end{bmatrix} + \begin{bmatrix} B_V \\ B_T \end{bmatrix} u + \begin{bmatrix} \bar{0} \\ -B_T \end{bmatrix} u_p + \begin{bmatrix} \bar{0} \\ C_T \end{bmatrix} \hat{g}, \tag{6}$$

where A_p and C_T are matrices of appropriate dimensions resulting from the reformulation of f_p. The vector $\hat{g}$ is the guideway irregularity vector which is the solution of the stochastic differential equation of Ito type (Appendix A) which, together with Eq. (6), are the vibration and tracking control plant equations forming the basis of the optimal control problem statements developed in this section.

In order to determine the nonzero elements of $\hat{F}_i$ and $\hat{L}_i$ that form the control u a decomposition of Eq. (6) will be performed. As indicated in the presentation of the decomposition of the VSSSPS in Section IV, Requirement Set II consists of demands on the means of the vehicle body state variables x_V, whereas Requirement Set III can be interpreted as a set of demands on the zero-mean state vector $\hat{x} \triangleq x - E(x)$. This makes it possible to define two control vectors, $\bar{u}$ and $\hat{u}$, to minimize independently the cost functionals resulting from Requirement Sets II and III, respectively. Substituting the relations

$$x = \begin{bmatrix} x_V \\ x_T \end{bmatrix} = \begin{bmatrix} \hat{x}_V \\ \hat{x}_T \end{bmatrix} + \begin{bmatrix} E(x_V) \\ E(x_T) \end{bmatrix}, \tag{7}$$

$$u \triangleq \bar{u} + \hat{u} \tag{8}$$

into Eq. (6) and taking the deterministic preload compensation vector u_p to be identical to $\bar{u}$ we obtain

$$\begin{bmatrix} \dot{\hat{x}}_V + \dfrac{dE(x_V)}{dt} \\ \dot{\hat{x}}_T + \dfrac{dE(x_T)}{dt} \end{bmatrix} = \begin{bmatrix} A_V & \bar{0} \\ \bar{0} & A_P \end{bmatrix} \begin{bmatrix} \hat{x}_V + E(x_V) \\ \hat{x}_T + E(x_T) \end{bmatrix}$$

$$+ \begin{bmatrix} B_V & \bar{0} \\ \bar{0} & B_T \end{bmatrix} \begin{bmatrix} \hat{u} + \bar{u} \\ \hat{u} \end{bmatrix} + \begin{bmatrix} \bar{0} \\ C_T \end{bmatrix} \hat{g}. \tag{9}$$

Define $\hat{u}$ to be a linear function of $x - E(x)$, and $\bar{u}$ to be a linear function of the mean $E(x)$. Then we have

$$E(\hat{u}) = 0,$$

$$E(\bar{u}) = \bar{u}. \tag{10}$$

Performing the expected value operation on Eq. (9) using (10) and the fact that

$$E(\dot{x}) = \frac{dE(x)}{dt} \tag{11}$$

we obtain

$$\begin{bmatrix} \dfrac{dE(x_V)}{dt} \\ \\ \dfrac{dE(x_T)}{dt} \end{bmatrix} = \begin{bmatrix} A_V & \bar{0} \\ \\ \bar{0} & A_P \end{bmatrix} \begin{bmatrix} E(x_V) \\ \\ E(x_T) \end{bmatrix} + \begin{bmatrix} B_V & \bar{0} \\ \\ \bar{0} & B_T \end{bmatrix} \begin{bmatrix} \bar{u} \\ \\ 0 \end{bmatrix} + \begin{bmatrix} \bar{0} \\ \\ C_T \end{bmatrix} \bar{0}. \tag{12}$$

The initial condition $x(0) = x_0$ can be taken such that $E(x_0) = 0$; hence Eq. (12) is equivalent to

$$\frac{dE(x_V)}{dt} = A_V E(x_V) + B_V \bar{u} \; ; \quad E\{x_V(0)\} = 0. \tag{13}$$

Furthermore, $E(x_T(t)) = 0$ for all $t \geq 0$, hence

$$\hat{x}_T(t) = x_T(t). \tag{14}$$

Subtracting Eq. (12) from (9) using (14) we have

$$\begin{bmatrix} \dot{\hat{x}}_V \\ \\ \dot{\hat{x}}_T \end{bmatrix} = \begin{bmatrix} A_V & 0 \\ \\ \bar{0} & A_P \end{bmatrix} \begin{bmatrix} \hat{x}_V \\ \\ x_T \end{bmatrix} + \begin{bmatrix} B_V \\ \\ B_T \end{bmatrix} \hat{u} + \begin{bmatrix} \bar{0} \\ \\ C_T \end{bmatrix} \hat{g}. \tag{15}$$

Defining

$$\hat{x} \triangleq \begin{bmatrix} \hat{x}_V \\ \\ x_T \end{bmatrix} \triangleq \begin{bmatrix} x_V - E(x_V) \\ \\ x_T \end{bmatrix}, \tag{16}$$

Eq. (15) can be rewritten as

$$\dot{\hat{x}} = A_s\hat{x} + B_s\hat{u} + C_s\hat{g}, \qquad \hat{x}(0) = \hat{x}_0,$$

$$E\{\hat{x}_0\} = E\{\hat{x}(t)\} = 0, \qquad E\{\hat{x}_0\hat{x}_0^T\} = S_0. \tag{17}$$

where A_s, B_s, and C_s in (17) are self defined from Eq. (15). To summarize, the decomposition presented in this section has resulted in the deterministic differential equation (13) and in the stochastic differential equation (17) with the stochastic process $\{\hat{g}_t, t \in T\}$ as a disturbance input. The controls $\bar{u}$ and $\hat{u}$ are independent by construction and by consideration of the assumptions stated in Section III. Therefore, they can be used to minimize independent cost functionals based on Requirement Sets II and III, respectively. Hence this decomposition has effectively removed the incompatibility of the vehicle suspension requirements contained in Sets II and III.

Note that the optimal preload-and-mass-control vectors have proven to be very useful in fulfilling Requirement Set I and linearizing the remaining suspension control synthesis problem at the first decomposition level (Fig. 1). It was possible to use advantageously the controllability of the primary preload again at the second decomposition level by introducing the preload compensation vector u_p to be identical to the deterministic control $\bar{u}$ to be calculated. This means that the reaction force of the deterministic control $\bar{u}$ on the trucks is canceled by an equal and opposite force generated by the primary preload. For this case the measurability of the guideway geometry acceleration vector (that is the acceleration experienced by the vehicle when it traverses the zero-roughness nominal guideway) can be shown.

B. Tracking Control Subsystem

1. Statement of the Deterministic Optimal Control Problem. By the decomposition presented in Section A, the secondary suspension force vector u composed of the nonzero elements of $\hat{F}_i$ and $\hat{L}_i$ consists of the controls $\bar{u}$ and $\hat{u}$ with the deterministic plant equation (13) and the stochastic differential equation (17), respectively. We shall now pose and solve the deterministic optimal control problem (DOCP) based on Eq. (13) and Requirement Set II of the VSSSPS.

PLANT EQUATION. The deterministic plant equation is given by

$$\frac{d\bar{x}_V(t)}{dt} = A_V\bar{x}_V(t) + B_V\bar{u}(t), \qquad \bar{x}_V(0) = 0, \tag{18}$$

where $\bar{x}_V \triangleq E(x_V)$ is the $(n_V \times 1)$ mean of vehicle body state vector x_V, $\bar{u}$ the $(m \times 1)$ deterministic control vector, and A_V, $(n_V \times n_V)$, B_V, and $(n_V \times m)$, are constant matrices.

ADMISSIBLE CONTROLS. As stated in Section III the elements of the vehicle state vector x are only measurable in certain linear combinations $E(y)$ of the state vector means which are given by the $(n_V \times 1)$ vector

$$E(y) = [T_1 \quad T_2]\begin{bmatrix} E(x_V) \\ E(x_T) \end{bmatrix} = T_1E(x_V), \tag{19}$$

where the second equality results from the fact that $E(x_T) = \bar{0}$. The matrix $[T_1 \quad T_2]$ is a submatrix of the access matrix D which selects the measurable combinations y_i out of the vector $[\hat{x}_t^T \quad \hat{g}_t^T]^T$ as will be shown in Section VII. A.1. Furthermore, D can be transformed without loss of generality so that T_1 is invertible. With this it is possible to define the set of admissible controls $\bar{u}$, $\psi(\bar{u})$ to be

$$\psi(\bar{u}) = \{\bar{u}(t); \quad \bar{u}(t) = \bar{K}(E\{y(t)\} - c(t)),$$

$$\|\bar{K}\| < \infty \}, \tag{20}$$

where $c(t)$ is a known $(n_V \times 1)$ command vector. The control $\bar{u}$ is seen to be bounded but not constrained in magnitude.

COMMAND VECTOR. The DOCP can be interpreted as a servomechanism problem where the control $\bar{u}$ forces given linear combinations of the components of the mean value of the vehicle-body state vector x_V, $E(x_V)$, to follow a command vector c which specifies values for the measurable linear combinations $T_1 E(x_V)_{\text{comm}}$ of the means of the vehicle body states. We have

$$c(t) = T_1 E\{x_V(t)\}_{\text{comm}}. \tag{21}$$

The elements of c have to be such that the demands of Requirement Set II are satisfied. This means that c consists of constants with the exception of the vehicle body roll command which must be calculated so that the lateral component of the resulting centrifugal and gravitational force is zero. The guideway of a high-speed tracked vehicle may be laid such that each mainline curve is connected to the straight track portion by transition spirals such that the lateral component of the resulting force increases linearly up to a constant value. Transition spiral are, however, sometimes omitted in track switches. The roll command for the vehicle body is assumed to consist, therefore, of arbitrary ramps and steps. All other commands are constant.

CONSTRAINTS. There are no displacement constraints for the tracking problem because they apply only to the vibration isolation. Instead of constraints the optimal trajectory $\bar{x}_V^0$ has to obey transient requirements with respect to damping and response time which are given in the VSSSPS. The response time requirement is based on the speed with which a vehicle traverses a typical mainline curve or a transition into a grade.

COST FUNCTIONAL. The cost functional for the DOCP is of quadratic form and can be formulated as

$$J(\bar{u}) = \frac{1}{2}\int_0^{T_f} ((E\{y(\tau)\} - c(\tau))^T Q_m (E\{y(\tau)\} - c(\tau))$$

$$+ \bar{u}(\tau)^T R_m \bar{u}(\tau))\ d\tau. \tag{22}$$

The final time T_f can be chosen to be infinity because of the assumed form of the command vector and the second-order vehicle equations of motion so that it is possible to obtain a time-invariant optimal control law. The tracking-weighting matrix Q_m and the control weighting matrix R_m are both symmetric and positive definite. The elements of Q_m and R_m will be selected in order to fulfill Requirement Set II.

2. Solution of the DOCP. The DOCP will be solved simply by reducing it to an infinite-time regulator problem. The solution for this problem is well known, therefore, only the result will be stated.

OPTIMAL DETERMINISTIC TRACKING CONTROL. The optimal tracking control $\bar{u}^0(t) \in \psi(\bar{u})$ that minimizes the cost functional (22) with the plant equation (18) is given by

$$\bar{u}^0(t) = -R_m^{-1}B_V^T\Pi_m T_1^{-1}[(T_1E\{x_V(t)\} + T_2E\{x_T(t)\})$$

$$- T_1E\{x_V(t)\}_{\text{comm}}], \tag{23}$$

where Π_m is the unique, constant, symmetric, and positive definite solution of

$$\bar{0} = \Pi_m A_V + A_V^T\Pi_m - \Pi_m B_V R_m^{-1} B_V^T \Pi_m + T_1^T Q_m T_1. \tag{24}$$

As shown in Eq. (14), $E(x_T) = 0$. It is, however, advantageous to include $E(x_T)$ in the mechanization of the tracking control in order to assure that residual external forces not canceled by the open-loop preload-and-mass control can be neutralized. The optimal tracking control system is uniformly asymptotically stable and is given by

$$\frac{dE(x_V)}{dt} = (A_V - B_V R_m^{-1} B_V^T \Pi_m)E(x_V) - B_V R_m^{-1} B_V^T \Pi_m T_1^{-1}(T_2E(x_T)$$

$$- T_1E(x_V)_{\text{comm}}), \qquad E\{x_V(0)\} = 0. \tag{25}$$

This concludes the solution of the DOCP. The determination of the weighting factors for the matrices R_m and Q_m will be performed so that the given response time and damping constraints are obeyed. A possible way is to maximize each position element of the optimal trajectory $\bar{x}_V^0 \triangleq E(x_V)^0$ by a scalar reference trajectory of a second-order system with response time and damping defined by the DOCP. This procedure leads to inequalities that the elements of the system matrix of the optimal closed-loop system (25) have to satisfy. This allows the selection of the elements of R_m and Q_m of the Riccati equation (24) [1].

C. Vibration Isolation Subsystem

1. Statement of the Stochastic Optimal Control Problem. Based on Eq. (17) and Requirement Set III a stochastic optimal control problem (SOCP) will be posed.

PLANT EQUATION. The first part of the plant equation for the SOCP is the stochastic differential equation (17) which was obtained as a result of the decomposition shown in Section VII. A. The disturbance input into Eq. (17), the stochastic process $\{\hat{g}_t,\ t \in T\}$, has been modeled as shown in Appendix A, from the given guideway spectral density as the stationary solution of the stochastic differential equation of Ito type

$$d\hat{g}_t = F_g \hat{g}_t\, dt + dv_t, \tag{26}$$

where $\hat{g}(0)$ is normal, $E(\hat{g}(0)) = 0$, $E\{\hat{g}(0)\hat{g}^T(0)\} = R_0$, $\{v_t,\ t \in T\}$ is a $(p \times 1)$ Wiener process vector with incremental covariance $C_g C_g^T\, dt$, $\hat{g}(0)$ and v_t are independent, and

$$C_g C_g^T = C_{g_1} C_{g_1}^T + C_{g_2} C_{g_2}^T + C_{g_1} C_{g_2}^T e^{\lambda F_g} + e^{\lambda F_g} C_{g_2} C_{g_1}^T.$$

Note that in stochastic equations subscripts t denote time dependence. Defining a new state vector

$$z_t = \begin{bmatrix} \hat{x}_t \\ \hat{g}_t \end{bmatrix}, \qquad ([n_V + n_T + p] \times 1) \tag{27}$$

with

$$\hat{x} = \begin{bmatrix} x - x_V^0 \\ x_T \end{bmatrix}, \qquad ([n_V + n_T] \times 1),$$

it is possible to write the stochastic plant equation as the augmented equation

$$dz_t = Az_t + B\hat{u}_t\, dt + C\, dv_t, \qquad E\{z_0 z_0^T\} = \hat{S}_0 = \begin{bmatrix} S_0 & \bar{0} \\ \bar{0} & R_0 \end{bmatrix} \tag{28}$$

with $z(0) = z_0$ normal with zero mean, R_0, S_0 symmetric and defined in Eqs. (26) and (17), z_0 and v_t independent, $\hat{u}_t$ the $(m \times 1)$ stochastic control, and

$$A = \begin{bmatrix} A_s & C_s \\ \bar{0} & F_g \end{bmatrix}, \qquad ([n_V + n_T + p] \times [n_V + n_T + p]),$$

$$B = \begin{bmatrix} B_s \\ \bar{0} \end{bmatrix}, \qquad ([n_V + n_T + p] \times m),$$

$$C = \begin{bmatrix} 0 \\ I \end{bmatrix}, \qquad ([n_V + n_T + p] \times p).$$

ADMISSIBLE CONTROLS. The state vector z_t is not completely measurable since the guideway irregularity vector $\hat{g}_t$ is not measurable from the moving vehicle. This means that the admissible controls cannot be functions of z_t but only functions of certain linear combinations of the components of z_t. Since the general structure of the suspension and the preload-and-mass control assures the measurability of the track-geometry acceleration we do not need preview information. Considering the assumptions stated in Section III we shall show that there exists at least a suboptimal control that leads to a stable system without the use of observer systems. We specify, therefore, instantaneous feedback.

The set of measurable components of z_t contains

(1) relative displacements between vehicle components and between trucks and guideway,

(2) relative velocities between the vehicle components, and

(3) relative velocities of the vehicle components relative to the intended guideway or moving track frame T obtained by integrating the difference between inertial acceleration of the particular component and the track geometry acceleration.

Hence the admissible controls are constrained in structure. The matrix which selects the measurable combinations y_i out of the elements of z_t is called the "access matrix" D such that

$$\begin{bmatrix} y_1(t) \\ \vdots \\ y_{n_V+n_T}(t) \end{bmatrix} = y_t = Dz_t, \qquad D([n_V + n_T] \times [n_V + n_T + p]),$$

where D has rank $(n_V + n_T)$. With this it is possible to define the set of admissible controls $\psi(\hat{u})$ as

$$\psi(\hat{u}) = \{\hat{u}_t : \hat{u} = \hat{K}Dz_t; \quad \|\hat{K}\| < \infty, \quad D \text{ given}\}. \tag{29}$$

Hence the set of admissible stochastic control $\psi(\hat{u})$ contains all linear time-invariant combinations of Dz_t such that $\hat{u}_t$

is bounded but not constrained in magnitude.

CONSTRAINTS. Requirement Set III contains the original Requirement (4) which stipulates that the vibration isolation of the vehicle body has to be performed within given displacement constraints. The vehicle component displacement for which constraints have been formulated are a subset of the elements of y_t selected by the constraint matrix T_c such that the vehicle component constraint vector y_c is given by

$$\begin{matrix} y_{c1}(t) \\ \vdots \\ y_{c\ell}(t) \end{matrix} \quad \triangleq \quad y_{ct} = T_c D z_t, \qquad T_c(\ell \times [n_V + n_T])$$

for $\ell < n_V + n_T$. Since the disturbance input into the plant equation of the stochastic optimal control problem is a Wiener process vector it is necessary to attach a probability P_i to each constraint. Since the plant equations are linear it suffices to select a probability so that at least the 3σ-value of the particular y_{ci} obeys the constraint. Hence it is possible to formulate the constraints as

$$y_{ct} = T_c D z_t \tag{30}$$

with elements $y_{ci}(t)$ such that

$$P_i\{|y_{ci}(t)| \leq a_i\} \geq 0.99865, \qquad i = 1,\ldots,\ell, \quad t \geq 0.$$

COST FUNCTIONAL. Based on Requirement Set III the cost functional for the stochastic optimal control problem has been selected to be

$$J(\hat{u}) = E\{\frac{1}{2}\int_0^{T_f} (\dot{\hat{x}}_\tau^T S \dot{\hat{x}}_\tau + z_\tau^T Q z_\tau + \hat{u}_\tau^T R \hat{u}_\tau)\, d\tau\} \tag{31}$$

with S, R, Q symmetric, R, $Q > 0$, and $S \geq 0$. The cost functional has been chosen to be quadratic because the resulting control is then necessarily linear and is calculable with a reasonable amount of computer time. The final time T_f is defined to be very large compared to the vehicle run time but finite. This allows us to chose T_f large enough to assure that the solutions of the Riccati equations to be derived in the next section are arbitrarily close to the steady-state solution. Therefore, it is possible to obtain time-invariant control laws that are easy to mechanize.

The elements of $\hat{x}$ are defined relative to the moving track frame T. Among the elements of $\dot{\hat{x}}$ are the coordinates of the relative vehicle body acceleration vector $\overset{\infty}{r}_{CV}$. Requirement (1) can be interpreted as minimization of $\overset{\infty}{r}_{CV}$ (a frequency weighting of $\overset{\infty}{r}_{CV}$ is possible by state augmentation but will not be considered here). Hence the cost functional minimizes the covariance of $\overset{\infty}{r}_{CV}$, $E\{(\overset{\infty}{r}_{CV} - E(\overset{\infty}{r}_{CV})) \cdot (\overset{\infty}{r}_{CV} - E(\overset{\infty}{r}_{CV}))^T\}$. The acceleration-selection matrix S selects the acceleration combinations to be minimized out of the vehicle state vector derivative $\dot{\hat{x}}$.

The second term in $J(\hat{u})$ is the displacement and velocity weighting term with the displacement and velocity weighting matrix Q. Here Q has been chosen to be positive definite and symmetric in order to assure stability for the optimal system with incomplete feedback.

Using Eq. (28), $\dot{\hat{x}}_t$ can be expressed as a function of z_t and $\hat{u}_t$. The cost functional $J(\hat{u})$ can, therefore, be written as

$$J(\hat{u}) = E\{\frac{1}{2}\int_0^{T_f} (z_\tau^T L z_\tau + \hat{u}_\tau^T M \hat{u}_\tau + z_\tau^T V \hat{u}_\tau + \hat{u}_\tau^T V^T z_\tau)\, d\tau\}, \tag{32}$$

in which

$$L \triangleq [A_s : C_s]^T S [A_s : C_s] + Q > 0,$$

$$M \triangleq B_s^T S B_s + R > 0,$$

$$V \triangleq [A_s : C_s]^T S B_s.$$

2. Optimal Control Using Complete Instantaneous State Feedback: Solution, Structure and Suboptimal Approximation

The SOCP stated in Section VII. A.1 will now be solved by applying the lemma stated in Appendix B. We have with the lemma the following identity:

$$E\{\frac{1}{2}\int_0^{T_f} (z_\tau^T L z_\tau + \hat{u}_\tau^T M \hat{u}_\tau + z_\tau V \hat{u}_\tau + \hat{u}_\tau^T V^T z_\tau)\, d\tau\}$$

$$= E\{\frac{1}{2} z(0)^T P(0) z(0) + \frac{1}{2}\int_0^{T_f} (\hat{u}_\tau + M^{-1}(V^T + B^T P(\tau)) z_\tau)^T$$

$$\times M(\hat{u}_\tau + M^{-1}(V^T + B^T P(\tau)) z_\tau)\, d\tau$$

$$+ \frac{1}{2}(\int_0^{T_f} dv^T C^T P(\tau) z_\tau + \int_0^{T_f} z_\tau^T P(\tau) C\, dv_\tau$$

$$+ \int_0^{T_f} \operatorname{tr} P(\tau) C C_g C_g^T C^T\, d\tau)\}. \tag{33}$$

Since M is positive definite there is a unique $\hat{u}^0$, if it exists, that minimizes (33). It is given by

$$\hat{u}_t^0 = -M^{-1}(V^T + B^T P(t))z_t. \tag{34}$$

The minimum cost is given by

$$J(\hat{u}^0) = E\{\tfrac{1}{2} z(0)^T P(0) z(0) + \tfrac{1}{2}\int_0^{T_f} dv_\tau^T C^T P(\tau) z_\tau + \tfrac{1}{2}(\int_0^{T_f} z_\tau^T P(\tau) C\, dv_\tau + \int_0^{T_f} \mathrm{tr}\, P(\tau) C C_g C_g^T C^T \, d\tau\}, \tag{35}$$

which can be evaluated to give

$$J(\hat{u}^0) = \frac{1}{2}\, \mathrm{tr}\, P(0)\hat{S}_0 + \frac{1}{2}\int_0^{T_f} \mathrm{tr}\, P(\tau) C C_g C_g^T C^T \, d\tau. \tag{36}$$

It can be shown that for the vehicle guideway system (28) whose uncontrollable guideway states are asymptotically stable, $P(t)$ is the unique and positive definite solution of

$$\frac{dP(t)}{dt} = -P(t)(A-BM^{-1}V^T) - (A-BM^{-1}V^T)^T P(t) + P(t)BM^{-1}B^T P(t) + VM^{-1}V^T - L; \qquad P(T_f) = \bar{0}. \tag{37}$$

Hence $\hat{u}_t^0$ exists. Furthermore, $\lim_{T_f \to \infty} P(t) = \Pi$ exists and is positive definite. The optimal closed-loop system is uniformly asymptotically stable and is given by

$$dz_t = A^0 z_t \, dt + C\, dv_t, \qquad E\{z(0)\} = 0, \qquad E\{z(0)z(0)^T\} = \hat{S}_0,$$
$$A^0 = A - BM^{-1}(V^T + B^T P(t)). \tag{38}$$

An analysis with respect to the structure of $\hat{u}_t^0$ is possible and the result can be stated as follows.

PROPOSITION. The structure of the optimal unconstrained vibration control $\hat{u}^0$ is given by

$$\hat{u}_t^0 = K_1\hat{x}_t + K_2\hat{g}_t + K_3\dot{\hat{x}}_t, \tag{39}$$

where $K_1 = -R^{-1}B_s^T\Pi_{11}$, $K_2 = -R^{-1}B_s^T\Pi_{12}$, $K_3 = -R^{-1}B_s^TS$. The control weighting matrix R and the acceleration selection matrix S are defined in Eq. (31), B_s is defined in Eq. (17) and

$$\Pi = \begin{bmatrix} \Pi_{11} & \Pi_{12} \\ \Pi_{12}^T & \Pi_{22} \end{bmatrix}$$

is the unique positive definite steady-state solution of the Riccati equation (37).

<u>Proof</u>. By Eq. (34) we have

$$-M\hat{u}_t^0 = (V^T + B^T\Pi)z_t.$$

Since $V = [A_s : C_s]^T SB_s$ (Eq. (32)) we obtain, adding and subtracting the term $B_s^TSB_su_t^0$,

$$(B_s^TSB_s - M)\hat{u}_t^0 = B_s^TS([A_s : C_s]z_t + B_s\hat{u}_t^0) + B^TP(t)z_t.$$

Hence, by Eq. (17),

$$(B_s^TSB_s - M)\hat{u}_t^0 = B_s^TS\dot{\hat{x}} + B^TP(t)z_t.$$

The final result follows with $M = B_s^T S B_s + R$ (Eq. (32)).

Equation (39) shows that the optimal vibration control using instantaneous state feedback consists of three basic parts. The $\hat{x}$ feedback term provides system stability; the $\hat{g}$ term is present because of the assumption of complete state feedback; the third term provides the acceleration feedback selected by S out of the state derivative vector $\dot{\hat{x}}$. It does *not* depend on the solution of the Riccati equation, hence the selection of the weighting factors in R and S only determine the acceleration feedback gains.

Relating this result to the SOCP we arrive at the following conclusions:

(1) The optimal vibration control $\hat{u}^0$ is time invariant only if $T_f \to \infty$.

(2) The optimal vibration control $\hat{u}^0$ requires measurability of the guideway irregularities.

The admissible controls $\hat{u}$ of the SOCP are such that they are time invariant and constrained in structure. Hence any $\hat{u} \in \psi(\hat{u})$ could at most be a suboptimal control for the case of finite time T_f and complete instantaneous state feedback.

Based on this consideration a suboptimal control will be proposed for $T_f \to \infty$ using the notion of the pseudoinverse D^I of the access matrix D. Since D has full rank we have

$$D^T = D^T[DD^T]^{-1}. \tag{40}$$

As stated in the SOCP only the combinations $y_t = Dz_t$ are measurable. Since D has no inverse we use the pseudoinverse to obtain the best possible solution for z_t, $\hat{z}_t$, with respect to the Euclidean norm to be

$$\hat{z}_t = D^I y_t. \tag{41}$$

Substituting (40) and (41) into (39) we obtain the following result.

SUBOPTIMAL STOCHASTIC CONTROL FOR THE SOCP CASE I. The following suboptimal time-invariant control case I is proposed as a mechanizable solution of the SOCP stated in Section VII. C.1:

$$\hat{u}_I^s(t) = -R^{-1}(B^T \Pi D^T [DD^T]^{-1} y_t + B_s^T S\hat{\dot{x}}_t), \tag{42}$$

where Π is the positive definite steady-state solution of Eq. (37). The suboptimal closed-loop system is given by

$$dz_t = A_I^s z_t\, dt + C\, dv_t, \qquad E\{z_0\} = 0, \qquad E\{z_0 z_0^T\} = \hat{S}_0,$$

$$A_I^s = A - BM^{-1}(B^T \Pi D^T [DD^T]^{-1} D + V^T). \tag{43}$$

Note that $\hat{u}_t^s$ can be rewritten as

$$u_I^s = -M^{-1}(B^T \Pi D^T [DD^T]^{-1} D + V^T) z. \tag{44}$$

Hence $u_I^s \notin \psi(\hat{u})$ if $V \neq 0$. This is caused by the fact that $\hat{u}_I^s$ utilizes the measurability of the vehicle-to-guideway accelerations contained in $\hat{\dot{x}}$. Hence the term $-M^{-1}V^T z$ need not be approximated by a linear combination of the measurements y_t.

The necessary and sufficient condition for uniform asymptotic stability for the zero-input system (43) can be shown to be the positive definiteness of the matrix Q^s:

$$Q^{S} = T_{I}[L + \Pi BM^{-1}B^{T}\Pi - VM^{-1}V^{T} - \Pi BM^{-1}B^{T}\Pi(I-\hat{D})$$

$$- (I-\hat{D})\Pi BM^{-1}B^{T}\Pi]T_{I}^{T}, \tag{45}$$

where $T_{I} = [I : \bar{0}]$, $([n_{V} + n_{T}] \times [n_{V} + n_{T} + p])$, I is the $([n_{V} + n_{T}] \times [n_{V} + n_{T}])$ identity matrix, and Π is the positive steady-state solution of Eq. (37).

3. Solution of the SOCP with Instantaneous State Feedback and Inaccessible States. By eliminating $\hat{u}$ from the cost functional (32) using (29) and utilizing the Ito differentiation rule the SOCP will be converted into an unconstrained matrix minimization problem [1]. Substituting $\hat{u} = \hat{K}Dz_{t}$ into (32) we obtain

$$J(\hat{u}) = E\{\frac{1}{2}\int_{0}^{T_f} z_{\tau}^{T}[L + D^{T}\hat{K}^{T}M\hat{K}D + D^{T}\hat{K}^{T}V^{T} + V\hat{K}D]z_{\tau}\, d\tau\}. \tag{46}$$

In order to assure that the optimal control $\hat{u}^{0}$ uses only instantaneous state feedback it is necessary to define a functional. Let this functional be the conditional expectation

$$V(t, z) = E\{\frac{1}{2}\int_{t}^{T_f} z_{\tau}^{T}[L + D^{T}\hat{K}^{T}M\hat{K}D + D^{T}\hat{K}^{T}V^{T} + V\hat{K}D]$$

$$z_{\tau}\, d\tau | z_{t} = z\}. \tag{47}$$

Note that by the properties of conditional expectation we have

$$E\{V(0, z_{0})\} = E\{E\{\frac{1}{2}\int_{0}^{T_f} z_{\tau}^{T}[L + D^{T}\hat{K}^{T}M\hat{K}D + D^{T}\hat{K}^{T}V^{T} + V\hat{K}D]$$

$$z_{\tau}\, d\tau | z_{0} = z_{0}\}\} = J(\hat{u}). \tag{48}$$

Since $V(t, z)$ fulfills the prerequisites for application of the Ito differentiation rule we obtain the following partial differential equation [2]:

$$\frac{\partial V}{\partial t} + \frac{1}{2} z_t^T [L + D^T \hat{K}^T M \hat{K} D + D^T \hat{K}^T V^T + V \hat{K} D] z_t$$

$$+ \frac{\partial V^T}{\partial z} (A + B\hat{K}D) z_t + \frac{1}{2} \operatorname{tr} \frac{\partial^2 V}{\partial z^2} C C_g C_g^T C^T = 0, \tag{49}$$

$$V(T_f, z) = 0.$$

Equation (49) can be recognized as the stochastic Hamilton-Jacobi equations of system (28) with cost (46). In fact, letting $C_g C_g^T = 0$, i.e., v_t a deterministic input, (49) reduces to the general form of the deterministic Hamilton-Jacobi equation with $\hat{u} = \hat{K} D z_t$. To solve Eqs. (49), let

$$V(t, z) = \frac{1}{2} z_t^T \hat{P}(t) z_t + p(t). \tag{50}$$

Substituting (50) into (49) we obtain differential equations for $\hat{P}(t)$ and $p(t)$:

$$\dot{\hat{P}}(t) = -\hat{P}(t)(A + B\hat{K}D) - (A + B\hat{K}D)^T \hat{P}(t)$$

$$- [L + D^T \hat{K}^T M \hat{K} D + D^T \hat{K}^T V^T + V \hat{K} D],$$

$$\hat{P}(t_f) = 0. \tag{51}$$

$$\dot{p}(t) = - \frac{1}{2} \operatorname{tr} \hat{P}(t) C C_g C_g^T C^T,$$

$$p(T_f) = 0. \tag{52}$$

By Eq. (48) we have

$$E\{V(0, z_0)\} = J(\hat{u}) = \frac{1}{2} E\{z_0^T \hat{P}(0) z_0 + 2p(0)\} \tag{53}$$

and evaluating Eq. (53) we obtain

$$J(\hat{u}) = \frac{1}{2} \operatorname{tr} \hat{P}(0)\hat{S}_0 + p(0). \tag{54}$$

Considering Eqs. (51), (52), and (54) we have thus succeeded in rewriting the SOCP as the following unconstrained matrix minimization problem: Given Eqs. (51) and (52) find $\hat{K}$, $\|\hat{K}\| < \infty$ (i.e., $\hat{K}$ is bounded but not constrained) which minimizes the cost (54).

The Athans matrix minimum principle [3] has been utilized to solve this matrix minimization problem and the following result has been obtained [1].

OPTIMAL STOCHASTIC CONTROL - INCOMPLETE STATE FEEDBACK. The optimal stochastic control $\hat{u}^0 \in \psi(\hat{u})$ that minimizes the cost function $J(\hat{u})$ of the SOCP is given, if it exists, by

$$u_t^0 = -M^{-1}[B^T\hat{P}(t) + V^T]\bar{S}(t)D^T[D\bar{S}(t)D^T]^{-1}Dz_t, \tag{55}$$

where $\hat{P}(t)$ and $\bar{S}(t)$ are solutions of the following two-point boundary value problem:

$$\dot{\hat{P}}(t) = -\hat{P}(t)(A + B\hat{K}^0D) - (A + B\hat{K}^0D)^T\hat{P}(t)$$

$$-[L + D^T(\hat{K}^0)^T M\hat{K}^0D + D^T(\hat{K}^0)^T V^T + V\hat{K}^0D],$$

$$\hat{P}(T_f) = 0. \tag{56}$$

$$\dot{\bar{S}}(t) = (A + B\hat{K}^0 D)\bar{S}(t) + \bar{S}(t)(A + B\hat{K}^0 D)^T + \frac{1}{2} CC_g C_g^T C^T,$$

$$\bar{S}(0) = \frac{1}{2}\hat{S}_0, \tag{57}$$

where

$$\hat{K}^0 = -M^{-1}[B^T\hat{P}(t) + V^T]\bar{S}(t)D^T[D\bar{S}(t)D^T]^{-1} \tag{58}$$

assuming $D\bar{S}(t)D^T$ is invertible. The minimum cost is given by

$$J(\hat{u}^0) = \frac{1}{2} \operatorname{tr} \hat{P}(0)\hat{S}_0 + \frac{1}{2}\int_0^{T_f} \operatorname{tr} \hat{P}(\tau)CC_g C_g^T C^T \, d\tau. \tag{59}$$

The second term of (59) has been obtained by solving Eq. (52) for $p(0)$. The optimal closed-loop system is given by

$$dz_t = \hat{A}^0 z_t \, dt + C \, dv_t, \qquad E\{z(0)\} = 0,$$

$$E\{z(0)z(0)^T\} = \hat{S}_0$$

$$\hat{A}^0 = A - BM^{-1}[B^T\hat{P}(t) + V^T]\bar{S}(t)D^T[D\bar{S}(t)D^T]^{-1}D. \tag{60}$$

Equations (56) and (57) represent a nonlinear two-point boundary value problem of high dimensionality. The solution of such formidable a problem on a computer is difficult and very time-consuming [3, 4]. However, the derivation in this section has resulted in an insight into the structure of the optimal control $\hat{u}^0$ with control structure constraints. Note that if $D = I$, the optimal control for incomplete state feedback reduces to the optimal control for complete state feedback, Eq. (34), as expected. Furthermore, the equation system (56) and (57) decomposes into two Riccati equations where (56) reduces to (37). Based on these insights a suboptimal control will be proposed.

It can be shown that the covariance $E\{z_t z_t^T\} = \hat{S}(t)$ of the optimal closed-loop system (60) is the solution of

$$\dot{\hat{S}}(t) = (A + B\hat{K}^0 D)\hat{S}(t) + \hat{S}(t)(A + B\hat{K}^0 D)^T + CC_g C_g^T C^T,$$

$$\hat{S}(0) = \hat{S}_0. \tag{61}$$

Comparing this with Eq. (57) we see that the costate matrix $\bar{S}(t)$ is related to the covariance of the optimal closed-loop system $\hat{S}(t)$ by

$$\hat{S}(t) = 2\bar{S}(t). \tag{62}$$

Furthermore, $\hat{K}^0$ can be rewritten as

$$\hat{K}^0 = -M^{-1}[B^T P(t) + V^T]\hat{S}(t)D^T[D\hat{S}(t)D^T]^{-1}. \tag{63}$$

Hence $\hat{u}_t^0$ depends explicitly on the covariance $\hat{S}$ of the optimal system.

McLane [3] has shown that there exists a suboptimal control that is based on the assumption that $\hat{S}$ is a given constant matrix such that $D\hat{S}D^T$ has an inverse. Adding and subtracting the term $B_s^T\bar{S}B_s u^0$ to Eq. (55) we obtain, using (63),

$$\hat{u}_t^0 = -R^{-1}[B^T\hat{P}(t)\hat{D}y_t + B_s^T S([A_s : C_s]\hat{D}Dz_t + B_s\hat{u}^0)], \tag{64}$$

where $\hat{D} = \hat{S}(t)D^T[D\hat{S}(t)D^T]^{-1}$. The term $[A_s : C_s]\hat{D}Dz_t + B_s\hat{u}^0$ can be recognized to be an estimate of the derivative $\dot{\hat{x}}$ of the state vector $\hat{x}$. Since the matrix S selects the measurable vehicle component accelerations relative to the "ideal" guideway of $\dot{\hat{x}}$, we conclude that we can replace the second term in Eq. (64) by $B_s^T S\dot{\hat{x}}_t$. We, therefore, propose the suboptimal control $\hat{u}^{s}_{II}$ to be

$$\hat{u}^{s}_{II}(t) = -R^{-1}[B^{T}\hat{P}(t)\hat{D}y_{t} + B^{T}_{s}S\dot{\hat{x}}_{t}]. \tag{65}$$

Since $\dot{\hat{x}} = [A_{s} : C_{s}]z_{t} + B_{s}\hat{u}^{s}_{II}(t)$ for the suboptimal system, we can rewrite Eq. (65) using the definition of V in Eq. (32):

$$\hat{u}^{s}_{II}(t) + -M^{-1}[B^{T}\hat{P}(t)\hat{D}D + V^{T}]z_{t}. \tag{66}$$

Having established the structure of $\hat{u}^{s}_{II}$ we proceed to choose a constant matrix $\hat{S}$ replacing $\hat{S}(t)$ in Eq. (63) based on the following engineering motivation: From the suspension synthesis problem statement we have an a priori knowledge of the form of the covariance of the optimal stochastic system because the Requirement Set II stipulates that the vehicle body acceleration must be minimized within the given displacement constraints of the vehicle. Considering the augmented state vector $z^{T} = [x^{T} \; g^{T}]$ we can utilize the fact that the average displacements $\alpha_{1}^{1/2}$ for the elements of the vehicle state $\hat{x}$ are much larger than the rms value $\alpha_{2}^{1/2}$ of the average displacements of the elements of the guideway irregularities $\hat{g}$ and assume the following form of $\hat{S}$:

$$\hat{S} = \begin{bmatrix} \alpha_{1}I_{n_{V}+n_{T}} & \bar{0} \\ & \\ \bar{0} & \alpha_{2}I_{p} \end{bmatrix}, \qquad \alpha_{1} > \alpha_{2} > 0, \tag{67}$$

where I_{i} is an i-dimensional identity matrix. With this we now propose the following form of the suboptimal control $\hat{u}^{s}_{II}$.

$$u^{s}_{II}(t) = -R^{-1}[B^{T}\tilde{P}(t)\hat{S}D^{T}[D\hat{S}D^{T}]^{-1}]y_{t} + B^{T}_{s}S\dot{\hat{x}}, \tag{68}$$

where $\hat{S}$ is given by Eq. (67) and $y_t = [I_{n_V+n_T} \quad D_3]z_t$ since the access matrix D can be taken without loss of generality to be

$$D = [I_{n_V+n_T} \quad D_3]. \tag{69}$$

Following the approach of McLane [3] we obtain the Riccati equation for $\tilde{P}(t)$ by substituting $\hat{K}^0 D$ by $-M^{-1}(B^T\tilde{P}(t)\tilde{D}D + V^T)$ in Eq. (56):

$$\begin{aligned}\dot{\tilde{P}}(t) &= -\tilde{P}(t)(A - BM^{-1}V^T) - (A - BM^{-1}V^T)^T\tilde{P}(t)\\ &\quad - L + VM^{-1}V^T - [\tilde{D}D]^T\tilde{P}(t)BM^{-1}B^T\tilde{P}(t)\tilde{D}D\\ &\quad + \tilde{P}(t)BM^{-1}B^T\tilde{P}(t)\tilde{D}D + [\tilde{D}D]^T\tilde{P}(t)BM^{-1}B^T\tilde{P}(t);\\ \tilde{P}(t_f) &= \bar{0},\end{aligned} \tag{70}$$

where $\tilde{D}D = \hat{S}D^T[D\hat{S}D^T]^{-1}D$. Equations (68) and (70) show that we need not choose $\hat{S}$ explicitly. It suffices to specify $\tilde{D}$.

With $\hat{S}$ given by (67) and D given by (69) we assume $\alpha_1 > \alpha_2$ such that

$$\hat{S}D^T[D\hat{S}D^T]^{-1}D \triangleq \tilde{D}D \cong \begin{bmatrix} I_{n_V+n_T} & D_3 \\ \bar{0} & \bar{0} \end{bmatrix} = \begin{bmatrix} D \\ \bar{0} \end{bmatrix}. \tag{71}$$

We obtain, therefore,

$$\tilde{D} = \hat{S}D^T[D\hat{S}D^T]^{-1} = \begin{bmatrix} I_{n_V+n_T} \\ \bar{0} \end{bmatrix}. \tag{72}$$

and can formulate a further result of this section.

SUBOPTIMAL STOCHASTIC CONTROL FOR THE SOCP CASE II. The following suboptimal time-invariant control $\hat{u}^{s}_{II}(t)$ is proposed as a mechanizable solution of the SOCP stated in Section VII. C.1.

$$\hat{u}^{s}_{II}(t) = -R^{-1}(B^{T}\tilde{\Pi}\tilde{D}y_t + B_s^{T}S\dot{\hat{x}}_t) \tag{73}$$

where

$$y_t = [I_{n_V+T} \quad D_3]z_t = Dz_t,$$

$$\tilde{D} = \begin{bmatrix} I_{n_V+n_T} \\ \\ \bar{0} \end{bmatrix}.$$

Then $\tilde{\Pi}$ exists and is the unique solution of

$$\begin{aligned} \bar{0} = & -\tilde{\Pi}(A - BM^{-1}V^{T}) - (A - BM^{-1}V^{T})^{T}\tilde{\Pi} \\ & - L + VM^{-1}V^{T} - D^{T}\tilde{D}^{T}\tilde{\Pi}BM^{-1}B^{T}\tilde{\Pi}\tilde{D}D \\ & + \tilde{\Pi}BM^{-1}B^{T}\tilde{\Pi}\tilde{D}D + D^{T}\tilde{D}^{T}\tilde{\Pi}BM^{-1}B^{T}\tilde{\Pi}. \end{aligned} \tag{74}$$

It can be shown that $\hat{u}^{s}_{II}(t)$ exists. The suboptimal closed-loop system is uniformly asymptotically stable (Appendix C) despite the use of incomplete state feedback and is given by

$$dz_t = A^{s}_{II}z_t\,dt + C\,dv_t, \qquad E\{z_0\} = 0, \qquad E\{z_0z_0^{T}\} = \hat{S}_0,$$

$$A^{s}_{II} = A - BM^{-1}[B^{T}\tilde{\Pi}\tilde{D}D + V^{T}]. \tag{75}$$

The covariance $S^{s}(t)$ of the suboptimal system (75) is the solution of

$$S^{s}(t) = A_{II}^{s}S^{s}(t) + S^{s}(t)A_{II}^{s})^{T} + CC_{g}C_{g}^{T}C^{T} \tag{76}$$

with initial condition $S^{s}(0) = \hat{S}_0$.

Comparing with Eq. (61) we see that the above equation is identical to Eq. (61) if $\hat{K}^0 D$ is substituted by $-M^{-1}(B^{T}\tilde{\Pi}\tilde{D}D + V^{T})$. Hence if $\lim_{t\to\infty} S^{s}(t) = \hat{S}$ such that $\hat{S}D[D\hat{S}D^{T}]^{-1} = \tilde{D}$ as given by Eq. (72), then $\hat{S}/2$ and $\tilde{\Pi}$ are steady-state solutions of the two-point boundary value problem (56) and (57), and $\hat{u}_{II}^{s}$ is identical to the optimal control $\hat{u}^0$ of the SOCP with control structure constraints.

The minimum cost for the original $\hat{x}$ system associated with the suboptimal control $\hat{u}_{II}^{s}$ can be calculated to be [3]

$$J(\hat{u}_{II}^{s}) = \frac{1}{2}\ \mathrm{tr}\ \tilde{\Pi}\hat{S}_0. \tag{77}$$

Using D as given in Eq. (69) we shall now compare $\hat{u}_{I}^{s}$ and $\hat{u}_{II}^{s}$. We note that $\tilde{\Pi}\tilde{D} = \Pi\tilde{D}$, where Π is the unique positive definite steady-state solution of Eq. (37). Hence the suboptimal controls differ only in the terms $D^{T}[DD^{T}]^{-1}$ and $\tilde{D}$. Taking $\tilde{S} = (I_{n_V+n_T} + p)\alpha$, we see that $D^{T}[DD^{T}]^{-1} = \tilde{S}D^{T}[D\tilde{S}D^{T}]^{-1}$

By Eq. (71) we have

$$\tilde{D} \cong \hat{S}D^{T}[D\hat{S}D^{T}]^{-1},$$

where $$\hat{S} = \begin{bmatrix} \alpha_2 I_{n_V+n_T} & \bar{0} \\ \bar{0} & \alpha_1 I_p \end{bmatrix}$$ and $\alpha_2 > \alpha_1$.

We arrive, therefore, at the conclusion that $\hat{u}_{I}^{s}$ differs from

$\hat{u}^s_{II}$ only in the form of the assumed covariance of the suboptimal closed-loop system! A surprising results indeed considering the completely different ways we have used to obtain $\hat{u}^s_I$ and $\hat{u}^s_{II}$. In motivating the choice of $\hat{S}$ for u^s_{II} we have stated that $\alpha_2 > \alpha_1$. Hence the case $\alpha_1 = \alpha_2$ as assumed for $\hat{u}^s_I$ is less likely to occur. Furthermore, utilization of u^s_I requires formally a verification of the stability of the suboptimal closed-loop system where $\hat{u}^s_{II}$ results in a uniformly asymptotically stable suboptimal system as is shown in Appendix C.

Certainly, the covariance of the optimal system will have a diagonal whose elements have values somewhere between α_1 and α_2, but to obtain the optimal control $\hat{u}^0_t$ necessary to calculate this covariance we must solve Eqs. (56) and (57). Based on the comparison between the two suboptimal controls we have derived, we arrive at the conclusion that the suboptimal control $\hat{u}^s$ of the SOCP is given by $\hat{u}^s_{II}$ and summarize the following result.

SUBOPTIMAL STOCHASTIC VIBRATION-ISOLATION CONTROL. The following suboptimal time-invariant control is proposed as a mechanizable solution of the SOCP stated in Section VII. C.1:

$$\hat{u}^s_t = -R^{-1}(B^T\tilde{\Pi}\tilde{D}y_t + B^T_s S\dot{\hat{x}}_t), \tag{78}$$

where $y_t = [I_{n_V+n_T} \quad D_3]z_t$ and $\tilde{D} \quad \begin{bmatrix} I_{n_V+n_T} \\ \bar{0} \end{bmatrix}$.

Then $\tilde{\Pi}$ exists and is the unique solution of Eq. (74) where the $([n_V+n_T] \times [n_V+n_T])$ partition Π_{11} of $\tilde{\Pi}$ is positive definite. The suboptimal stochastic closed-loop system is uniformly asymptotically stable and is given by Eq. (75). The covariance

$S^s(t)$ and the minimum cost are given by Eqs. (76) and (77), respectively.

The solution $\tilde{\Pi}$ of Eq. (74) depends on the choice of weighting factors of the matrices Q, S, and R which are contained in the expressions for the matrices L, M, and V in Eq. (32). One possible way is to use the covariance $S^s(t)$ of the suboptimal system (75): The SOCP states that the elements of the vehicle component constraint vector y_c have to obey the displacement constraints with a given probability P as defined in Eq. (30). By the decomposition of the vehicle equation (Section VII. A) we have that $E\{y_c(t)\} = 0$. Hence we can interpret Eq. (30) as a specification of the maximum covariance of each element y_{ci} and can write, assuming Gaussian distribution,

$$E\{y_{c_i}^2(t)\} \leq \frac{a_i^2}{9} \triangleq k_i \tag{79}$$

Substituting Eq. (30) into Eq. (79) we obtain the following condition:

$$\text{diag } T_c D S^s(t) D^T T_c^T \leq \begin{bmatrix} k_1 & & 0 \\ & \ddots & \\ 0 & & k_\ell \end{bmatrix}, \qquad t \geq 0 , \tag{80}$$

where D must be in the form given in the SOCP.

Based on inequality (80) a procedure can be developed by which the elements of Q, R, and S can be found for which is $\tilde{\Pi}$ is such that $S^s(t)$ obeys the inequality. The vibration control $\hat{u}^s$, therefore, fulfills Requirement Set III.

At this point it should be noted that the displacements constraints given in the VSSSPS are based on preliminary estimates. It is, therefore, possible that they are incompatible. For instance, if the displacement constraints for the secondary suspension are too small such that obeying these constraints would result in a violation of the truck-to-guideway displacement constraints, then the displacement specification must be redefined until a compatible set has been obtained.

VIII. CALCULATION OF THE SYNTHESIZED SECONDARY SUSPENSION FORCES AND TORQUES.

It has been shown that the zero-input deterministic tracking control system and the suboptimal stochastic vibration control system are uniformly asymptotically stable. The controls $\bar{u}^0$ and $\hat{u}^s$ have time-invariant gains and are constrained to use only measurable combinations of states. By Eq. (8) we have

$$u^s = \bar{u}^0 + \hat{u}^s, \tag{81}$$

where as defined in Eq. (5) the elements of $\hat{u}^s$ are defining the nonzero elements of the secondary suspension force vectors $\hat{F}_i$ and the secondary suspension roll torques $\hat{L}_i$.

Using Eqs. (1)-(4) together with the optimal preload-and-mass control vectors we are now able to specify completely the secondary suspension force vector F_i, the secondary suspension roll torque vector L_i, and the controllable preloads F_{pi} and L_{pi} and have, therefore, solved the suspension synthesis problem.

IX. AN EXAMPLE

In order to demonstrate the main advantage of the synthesized suspension structure--the removal of the incompatibility between vibration isolation and guideway tracking--an example will be presented [1].

Given the vehicle configuration shown in Fig. 3, which follows a guideway with arbitrary vertical curves, synthesize a levitation system that minimizes the vehicle body acceleration and tracks the guideway. The vehicle configuration parameters are:

vehicle body mass	$m_c = 207 \dfrac{\text{ft-lb sec}^2}{\text{in.}}$,
vehicle body pitch inertia	$I_p = 1.73 \times 10^6$ ft-lb sec^2 ft,
truck mass	$m_i = 25.9 \dfrac{\text{ft-lb sec}^2}{\text{in.}}$, $i=1,2$,
primary spring rate	$k = 50{,}000 \dfrac{\text{ft-lb}}{\text{in.}}$,
distance truck centerline to vehicle body centerline	$a = 35$ ft,
vehicle speed	$v = 300$ mph.
angular rotation about center of gravity	θ_p = radians

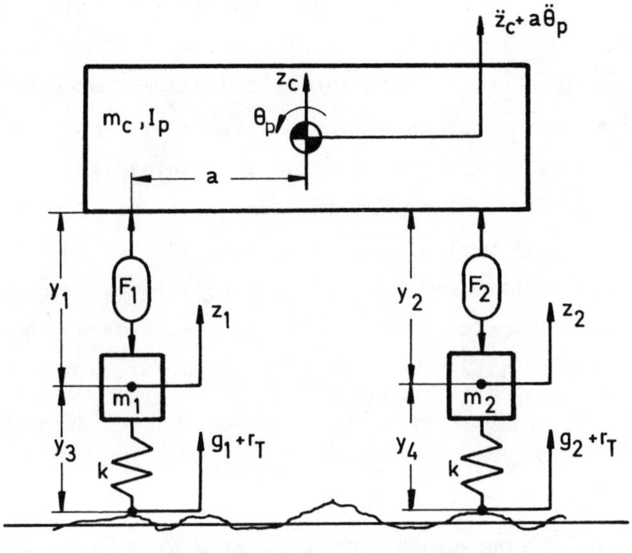

Fig. 3. Test model configuration and definition of generalized coordinates and secondary suspension forces.

Constraints

The following displacement constraints must be obeyed with a probability $P \geq 0.99865$:

secondary suspension displacements	$\lvert y_1 \rvert < 1.0$ in.,
	$\lvert y_2 \rvert < 1.0$ in.,
primary air gap variations	$\lvert y_3 \rvert < 0.25$ in.,
	$\lvert y_4 \rvert < 0.25$ in.,

deterministic constraints:

tracking response time $\quad T_r < T_{r_{max}} = 0.25$ sec

with damping factor

$\xi \geq 0.7$.

In order to keep the dimensionality of the example low only the levitation part of the overall vehicle suspension will be calculated since this test model is sufficient to demonstrate the superior performance of the synthesized suspension when compared with conventional configurations. Furthermore, applying the methods of Section VII to determine the preload control for the test model is straightforward and does not yield any new insight, hence the optimal preload-and-mass control is assumed to exist. We, therefore, start immediately with the second decomposition of the vehicle synthesis problem. The displacement constraints and the response time $T_{r_{max}}$ have been purposely chosen to be very tight in order to show that the new suspension structure is performing very well even under these severe constraints.

For complete formulations of the control problems we need to determine the measurable states. From Fig. 3, we can identify the following measurable combinations of states y_i:

$$y_1 = z_c - a\theta_p - z_1, \qquad y_2 = z_c + a\theta_p - z_2,$$

$$y_3 = z_1 - g_1, \qquad y_4 = z_2 - g_2,$$

$$y_5 = \mathring{z}_c - a\dot{\theta}_p, \qquad y_6 = \mathring{z}_c + a\dot{\theta}_p,$$

$$y_7 = \mathring{z}_1, \qquad y_8 = \mathring{z}_2.$$

Note y_7 and y_8 are available since the optimal preload-and-mass control allows determination of the guideway acceleration vector $\ddot{r}_T^T = [0 \;\; 0 \;\; \ddot{r}_{T_z}]$, hence

$$\mathring{z}_i(t) = \int_0^t (\ddot{z}_i(\tau) - \ddot{r}_{T_z}(\tau))\, d\tau, \qquad i = 1, 2.$$

Two digital computer programs have been written in order to solve for the optimal deterministic tracking control $\bar{u}^0$ and the suboptimal stochastic vibration isolation control $\hat{u}_I^s$. The latter has been chosen in order to demonstrate that the stability condition given in Eq. (45) holds.

The condition of stability, Eq. (45), $Q^s > 0$, has been checked using an eigenvalue routine for real symmetric matrices. All eigenvalues were found to be positive, hence the suboptimal vibration isolation system is uniformly asymptotically stable for zero input. Furthermore, the elements of S and Q in the stochastic Riccati equation have been chosen such that

$$E\{y_i^2(t)\} \le \begin{cases} 0.0932 \text{ in.}^2, & i = 1, 2, \\ 0.00675 \text{ in.}^2, & i = 3, 4. \end{cases}$$

Calculating the respective 3σ-values we see that

$$|y_1| = |y_2| = 0.92 < 1.0,$$

$$|y_4| = |y_5| = 0.248 < 0.25.$$

Hence the displacement constraints given in the problem statement are obeyed.

ANALOG COMPUTER TEST RUNS. With the calculation of the suspension forces

$$F_1 = \bar{F}_1 + \hat{F}_i, \qquad F_2 = \bar{F}_2 + \hat{F}_2,$$

where $\bar{F}_1$ and $\bar{F}_2$ are the elements of $\bar{u}^0$, and $\hat{F}_1$ and $\hat{F}_2$ are the elements of $\hat{u}_I^s$ we have synthesized a levitation system which fulfills the requirements of the problem statement of the given test model and, therefore, solves the suspension synthesis problem.

It is, however, very useful to carry the suspension design process one step further and perform an analog computer simulation of the test model and the synthesized suspension including the effect of the $E(y)$ sensors in order to show the following:

(1) effect of time constants of the $E(y)$ sensors on suspension performance, and

(2) comparison of synthesized suspension performance with conventional configurations.

A. $E(y)$ Sensors

As will be seen in Appendix A, the guideway irregularities can be considered to be ergodic stationary processes. Since the vehicle model is linear one possibility is to use time-averaging to obtain the means of the measurable linear combinations of state variables which are needed for the tracking control. Furthermore, it is well known that time-averaging of a signal y over an averaging time T_a is equivalent to passing y through a first-order lag with a time constant $\tau = T_a/2$. We have, therefore, obtained a particular simple way to obtain the expected values.

For the suspension synthesis problem we can choose the time constant τ of the expected value sensor based on the fact that all frequencies below τ^{-1} are not attenuated. Considering the one-sided guideway irregularity power spectral density model in the time domain $G^*(\omega)$, we can define τ to be larger than or equal to $1/\omega_u$, where ω_u is given by

$$a_{rms}^2 = \int_0^{\omega_u} \omega^4 G^*(\omega)\, d\omega.$$

Hence given the guideway irregularity PSD $G(\omega)$ and the maximum admissible root mean square acceleration a_{rms} caused by the irregularities which result in vehicle body accelerations less than the specified comfort level we can arrive at a suitable value of τ. In this particular example the velocity mean sensor had a destabilizing effect due to the large time constant chosen. By simulation on the analog computer it was found that this effect can be canceled by an increase of the gains for the feedbacks associated with $\hat{y}_5$ and $\hat{y}_6$. All other gains remained unchanged. As will be shown this gain change did not significantly alter the vibration-isolation characteristics of $\hat{F}_1$ and $\hat{F}_2$. This augures well for the use of the suspension synthesis procedure described combined with an assumption-checking analog computer simulation for high-speed vehicle suspension design.

B. Comparison with Conventional Suspension Configurations. In this section three analog computer runs will be presented which demonstrate how effective the new suspension structure is. If we consider the optimal deterministic and stochastic control problems separately, we conclude that each of them can be an optimal control problem statement for a suspension system. The only difference between the two problem statements is that one problem emphasizes tracking while the other emphasizes vibration isolation. Solving the respective optimal control problems we obtain two suspension configurations--both are using the same vector of measurable states--which have suspension force parameters identical to those obtained for the tracking and vibration control, respectively. These two vehicle suspension configurations will be denoted "the conventional suspension configurations". We have, therefore, applied optimal control theory to the suspension synthesis problem in the way the papers reviewed in a literature survey have used. Since each control operates on the same vector of measurable states we know that guideway tracking and vehicle body vibration isolation are incompatible and that either one can be emphasized only at the expense of the other by a suitable choice of the cost functional weighting factors. Hence, despite the use of optimal control theory we cannot remove the incompatibility between guideway tracking and vibration isolation without applying the decomposition presented.

For all three runs in the same guideway spectral density has been used. In order to better demonstrate the difference in performance between the three suspension systems, a test guideway roughness PSD has been assumed which results in smaller displacements of the secondary suspension:

$$G(\nu) = \frac{0.6 \times 10^{-4}}{\nu^2} \quad \frac{\text{in.}^2}{\text{rad/ft}}, \qquad \nu \geq \frac{2\pi}{4400} \quad \frac{\text{rad}}{\text{ft}}.$$

The analog computer runs I-III are shown in Figs. 4-6. The variables shown are identified in Fig. 3.

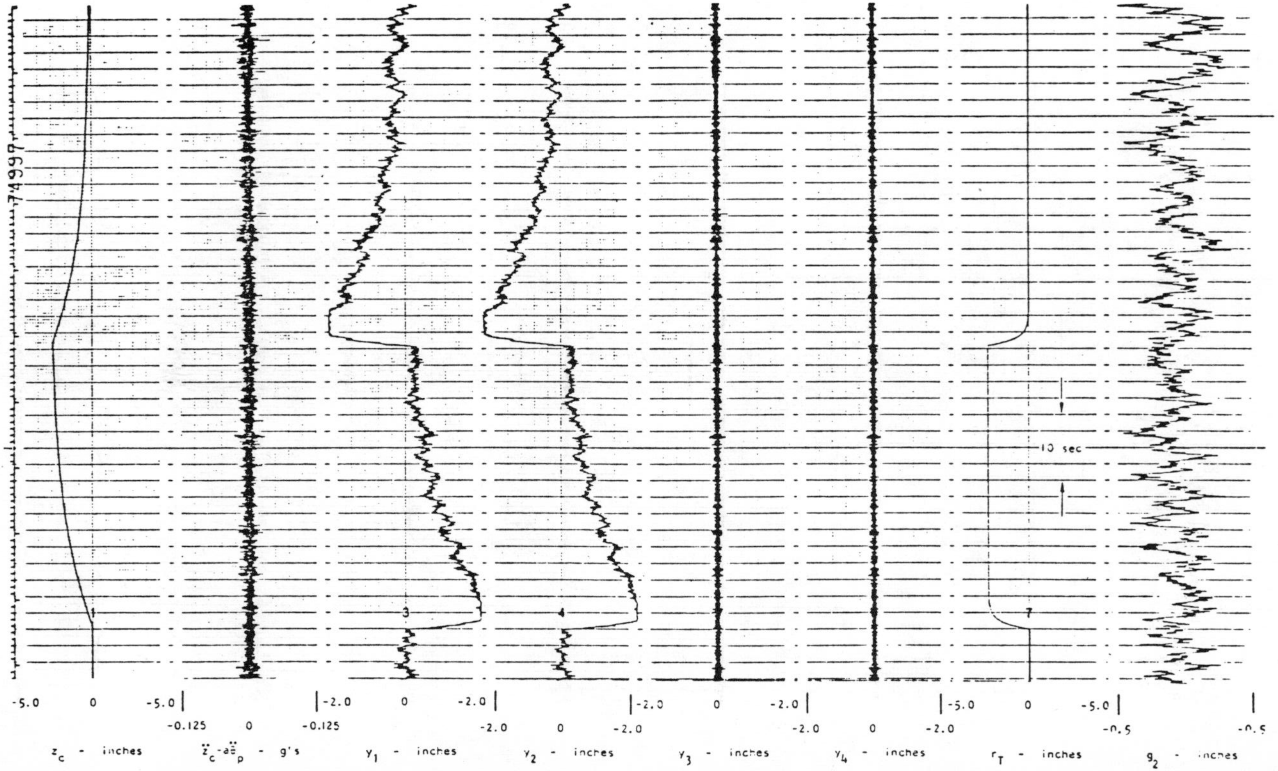

Fig. 4. Analog computer test run I: Conventional optimized suspension emphasizing vibration isolation.

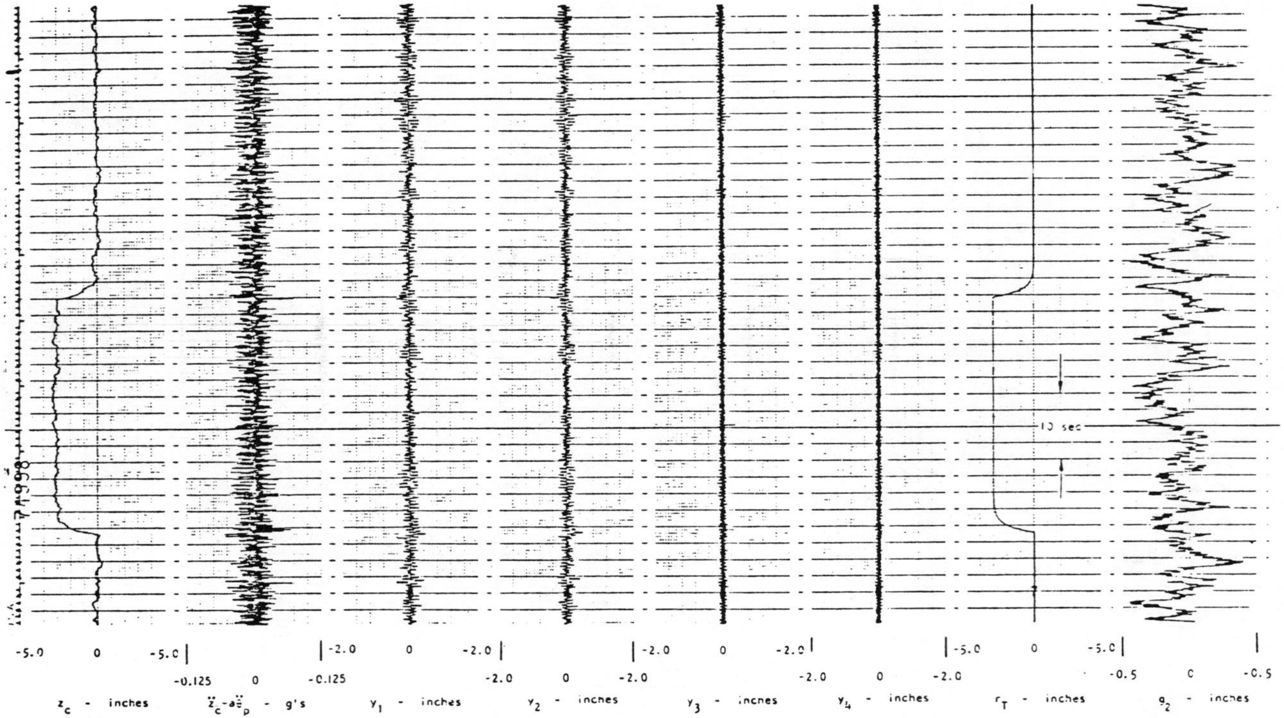

Fig. 5. Analog computer test run II: Conventional optimized suspension emphasizing guideway tracking.

RUN I: (Fig. 4). The vehicle suspension system under consideration is the conventional suspension configuration emphasizing vehicle body vibration isolation. Indeed, as seen in trace 2 the rms value of the vehicle body acceleration relative to the track frame T is $0.011g$ which is a very good vibration isolation performance. However, the resulting suspension is so soft that it cannot follow the track geometry change shown in trace 7 without a large deflection of the secondary suspension as shown in traces 3 and 4. We see that in order to follow a lagged guideway step with a height of 3 in. at a speed of 300 mph, the suspension squats more than 2 in.!

RUN II: (Fig. 5). This is a test run for the conventional vehicle configuration emphasizing guideway tracking. We see that it follows the track geometry change very well but at the expense of ride quality. The measured rms value of the relative vehicle body acceleration is $0.024g$ or more than twice the value obtained for the configuration of Run I.

RUN III: (Fig. 6). The suspension synthesized for the test model by the methods developed here with the $E(y)$ sensors as described in Section IX. A has been subjected to the same conditions as the two conventional configurations. It is seen that the rms value of the relative vehicle body acceleration is $0.0115g$ which is insignificantly larger than measured for the conventional suspension configuration of Run I. The increase is caused by the mechanization of the $E(\cdot)$ operation. At the same time, however, the synthesized suspension follows the track geometry change so well that it is difficult to pick the residual squat out of the displacement traces--a far cry from the 2-in. deflection seen in Run I. Note that in Run I no $E(y)$ sensors are present which manifests itself in the smooth trace of z_c. For the synthesized suspension the $E(y)$ sensors are mechanized. The effect of the low-frequency guideway irregularities on the

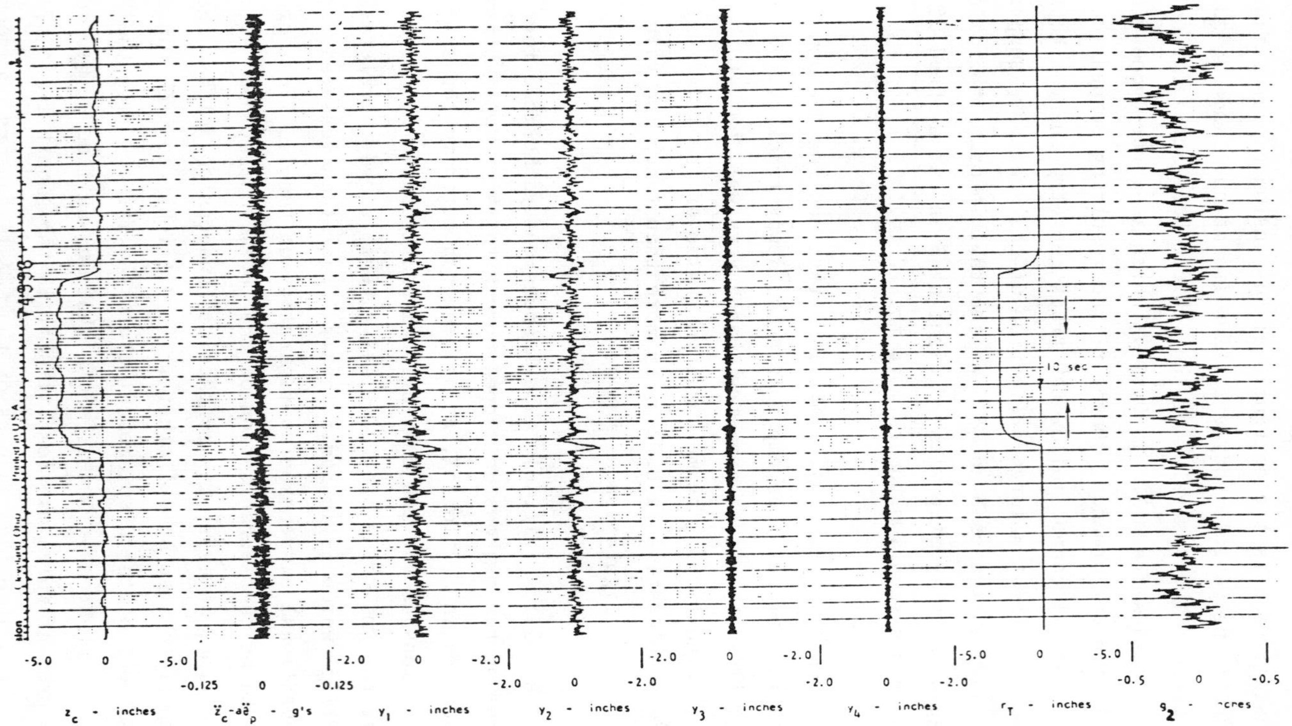

Fig. 6. Analog computer test run III: Synthesized suspension.

motion of the vehicle body is seen in trace 1. However, the acceleration traces of Runs I and II are nearly identical thus demonstrating the negligible effect of the $E(y)$ sensors on the vehicle body acceleration.

Summarizing the results of the digital computer calculations and analog computer simulation it can be stated that it is possible to synthesize and mechanize a suspension for which guideway tracking and vehicle body vibration isolation are not incompatible.

X. SUMMARY AND CONCLUSION

This chapter was concerned with the problem of synthesizing a suspension system for a high-speed tracked vehicle. Using a new decomposition technique it was shown that there exists an optimal structure that a suspension must have in order to fulfill all requirements. It was shown that the derived suspension structure is such that the incompatibility of vibration isolation and guideway tracking under the influence of external forces which is present for all suspensions synthesized so far--whether optimal control approaches have been used or not--has been removed.

The new suspension structure features three independent parts: the nonlinear and time-varying preload-and-mass control operating on external forces and vehicle body accelerations, the linear time-invariant tracking control operating onthe means of certain vehicle states, and the linear time-invariant vibration control which operates on the difference between the actual vehicle state vector and its mean.

It can be concluded that the application of the principles and methods of advanced dynamics, deterministic and stochastic optimal control theory, and applied mathematics to the problem of synthesizing a high-speed vehicle suspension was successful and has resulted in the identification of a new secondary suspension structure that is proposed as a starting point in the development and design of a high-speed vehicle suspension system. Augmented by a thorough system simulation and hardware tests for verification of all assumptions made, and implementation of the sensor and actuator dynamics the described synthesis procedure will lead to an effective suspension system.

REFERENCES

1. C.R. GUENTHER, "Application of Stochastic Optimal Control Theory to High-Speed Vehicle Suspension Synthesis", Ph.D. Thesis, University of California, Los Angeles.
2. K.J. ASTRÖM, "Introduction to Stochastic Control Theory". Academic Press, New York, 1965.
3. P.J. McLANE, Internat. J. Control, 13 (2), 383-396 (1971).
4. R. KOSUT, IEEE Trans. Automatic Control, AC-15 (5), 557-563 (1970).
5. M. ATHANS and P.L. FALB, "Optimal Control". McGraw-Hill, New York, 1966.

APPENDIX A: GUIDEWAY IRREGULARITY DESCRIPTION

Given the guideway irregularity g represented by a stationary stochastic process in the distance domain $\{g_s, s \in F\}$, where F is the length of the guideway, with a measured single-sided spatial spectral density (SD)

$$G_{gg}(\nu) = \frac{k}{\nu^2} ; \qquad \nu \geq \nu_0 \tag{A.1}$$

we generate a SD model in order to convert Eq. (A.1) into a stochastic differential equation in the distance domain.

The SD model must (1) contain all positive spatial frequencies ν, (2) be as close as possible to the measured SD for $\nu > \nu_0$, and (3) result in the same rms value as calculated for the SD given in Eq. (A.1). The SD model $G^*_{gg}(\nu)$ is given by the single sided spatial spectral density [1]

$$G^*_{gg}(\nu) = \frac{k}{((\pi\nu_0/2)^2 + \nu^2)}, \quad \nu \geq 0. \tag{A.2}$$

The random variable g_s of the stochastic process $\{g_s, s \in F\}$ characterized by $G^*_{gg}(\nu)$ is the stationary solution of the following linear stochastic differential equation of Ito type in the distance domain:

$$dg_s = -\frac{1}{2}\pi\nu_0\, g_s\, ds + (\pi\nu)^{\frac{1}{2}}\, d\eta_s. \tag{A.3}$$

In order to assure stationarity of $\{g_s, s \in F\}$, the initial state $g(0) \triangleq g_0$ must be Gaussian with $E\{g(0)\} = 0$ and covariance $E\{g^2(0)\} = k/\nu_0$. The driving function of the equation $\{\eta_s, s \in F\}$ is a Wiener process with unity incremental covariance, and η_s and g_0 are independent. Subscript s denotes the distance dependence.

The notion of a Wiener process can very conveniently be used to allow an implementation of the distance L between the truck centers. The truck length itself is conservatively taken to be zero; nonzero truck length can be taken into account by state augmentation. Let the stationary process $\{g_s, s \in F\}$ hit the rear truck of the vehicle at $s = s_0$; then it would have hit the front truck at $s = s_0 + L$. We can, therefore, define the stochastic process $\{h_s, s \in F\}$ to be the stationary solution of Eq. (A.3) with $d\eta_s$ replaced by $d\eta_{s+L}$. Hence the random variable h_s hitting the front truck equals the random variable g_{s+L} where g_s is just hitting the rear truck. In other words, h_s and g_s are readouts of the same irregularity trace separated by the length L.

Since the vehicle equations of motion are written with

respect to the track frame T, which moves along the intended guideway with constant velocity v, the random variables h_s and g_s must be transformed from the distance domain to the time domain such that h_t and g_t are functions of time with respect to the moving vehicle and are the stationary solutions of the following stochastic differential equations of Ito type in the time domain:

$$dh_t = -\frac{1}{2}\pi\nu_0 v h_t\, dt + (\pi k v)^{\frac{1}{2}}\, dw_{t+\lambda}, \tag{A.4}$$

$$dg_t = -\frac{1}{2}\pi\nu_0 v g_t\, dt + (\pi k v)^{\frac{1}{2}}\, dw_t, \tag{A.5}$$

with initial conditions $E\{h(0)\} = E\{g(0)\} = 0$, $E\{h^2(0)\} = E\{g^2(0)\} = k/\nu_0$, in which $\{w_t,\ t \in T\}$ is a Wiener process with unity incremental covariance, w_t is independent of $h(0)$ and $g(0)$ and $\lambda = L/v$. It is seen that the statistics of h and g are invariant with respect to the performed transformation.

Denoting the guideway irregularities acting on the front truck h_i, $i = 1,\ldots,3$, and the irregularities acting on the rear truck g_i, $i = 1,\ldots,3$, we can define the guideway irregularity vector $\hat{g}_t$ in the time domain for a continuously supported guideway by

$$\hat{g}_t^T \triangleq [g_t^T \quad h_t^T]. \tag{A.6}$$

For the elements of g_t, index 1 denotes vertical, index 2 lateral irregularity of the guideway, and index 3 the roll input generated by the vertical irregularities of the left and right rail. For this definition, the elements of g_t and h_t are independent, respectively.

For constant vehicle speed v the stochastic process $\{\hat{g}_t,\ t \in T\}$ is the stationary solution of the stochastic differential equation of Ito type

$$d\hat{g}_t = F_{\mathrm{g}}\hat{g}_t \, dt \quad C_{\mathrm{g}1} \, dw_t + C_{\mathrm{g}2} \, dw_{t+\lambda}; \tag{A.7}$$

$$F_{\mathrm{g}} = \begin{bmatrix} F_1 & \bar{0} \\ \bar{0} & F_1 \end{bmatrix}, \quad C_{\mathrm{g}1} = \begin{bmatrix} C_1 \\ \bar{0} \end{bmatrix}, \quad C_{\mathrm{g}2} = \begin{bmatrix} \bar{0} \\ C_1 \end{bmatrix},$$

where F_1 and C_1 are 3×3 diagonal matrices with elements determined by Eqs. (A.4) and (A.5). The initial vector $\hat{g}_0 \triangleq \hat{g}(0)$ is Gaussian with zero mean and covariance $E\{\hat{g}_0\hat{g}_0^{\mathrm{T}}\}$, the Wiener process $\{w_t, \ t \in T\}$ is an independent 3×1 vector with incremental covariance $I \, dt$, w_t and $\hat{g}_0$ are independent.

Equation (A.7) is the mathematical model of the guideway irregularities employing the notion of a time-shifted Wiener process. If $t > \lambda$, it is possible, however, to replace Eq. (A.7) by [1]

$$d\hat{g}_t = F_{\mathrm{g}}g_t \, dt + dv_t, \tag{A.8}$$

where $\{v_t, \ t \in T\}$ is a Wiener process 6×1 vector with incremental covariance $C_{\mathrm{g}}C_{\mathrm{g}}^{\mathrm{T}} \, dt$, where $C_{\mathrm{g}}C_{\mathrm{g}}^{\mathrm{T}}$ is given by

$$C_{\mathrm{g}}C_{\mathrm{g}}^{\mathrm{T}} = C_{\mathrm{g}1}C_{\mathrm{g}1}^{\mathrm{T}} + C_{\mathrm{g}2}C_{\mathrm{g}2}^{\mathrm{T}} + C_{\mathrm{g}1}C_{\mathrm{g}2}^{\mathrm{T}}e^{\lambda \hat{F}_{\mathrm{g}}} + e^{\lambda \hat{F}_{\mathrm{g}}}C_{\mathrm{g}2}C_{\mathrm{g}1}^{\mathrm{T}}. \tag{A.9}$$

Since F_{g} is diagonal, $e^{\lambda \hat{F}_{\mathrm{g}}}$ is easily calculatable.

Equation (A.9) has been obtained by calculating the differential equation of the covariance $E\{\hat{g}_t\hat{g}_t^{\mathrm{T}}\}$ of $\hat{g}_t$ based on Eq. (A.7), repeating the procedure for the equation $d\hat{g}_t = F_{\mathrm{g}}\hat{g}_t \, dt + C_{\mathrm{g}} \, dv_t$ with the same initial conditions as given for Eq. (A.5) but with $\{v_t, \ t \in T\}$ to be a Wiener process with incremental covariance $I\,dt$, and equating the respective terms in both covariance differential equations.

APPENDIX B. A USEFUL LEMMA.

LEMMA 3. Given the stochastic differential equation

$$dz_t = Az_t\,dt + Bu_t\,dt + C\,dv_t \tag{B.1}$$

where z is an $(n \times 1)$ state vector, u an $(m \times 1)$ control vector, $\{v_t, t \in T\}$ a $(p \times 1)$ Wiener process vector with incremental covariance R_0, z_0 normal with zero mean and covariance $\hat{S}_0$, z_0 and v_t independent, and A, B, and C are constant matrices of appropriate dimensions. Let z_t be a solution of (B.1) for a given bounded u_t. Then consider

$$\dot{P}(t) = -P(t)(A - BM^{-1}V^{\mathrm{T}}) - (A - BM^{-1}V^{\mathrm{T}})^{\mathrm{T}}\,P(t)$$

$$+ P(t)BM^{-1}B^{\mathrm{T}}P(t) + VM^{-1}V^{\mathrm{T}} - L, \quad P(T_{\mathrm{f}}) = K. \tag{B.2}$$

Assume that the solution $P(t)$ exists and is at least nonnegative definite. Then we have the following identity:

$$\frac{1}{2}\,z(T_{\mathrm{f}})^{\mathrm{T}}Kz(T_{\mathrm{f}}) + \frac{1}{2}\int_0^{\hat{T}_{\mathrm{f}}} (z_\tau^{\mathrm{T}}Lz_\tau + u_\tau^{\mathrm{T}}Mu_\tau + z_\tau^{\mathrm{T}}Vu_\tau + u_\tau^{\mathrm{T}}V^{\mathrm{T}}z_\tau)\,d\tau$$

$$= \frac{1}{2}\,z(0)P(0)z(0) + \frac{1}{2}\int_0^{\hat{T}_{\mathrm{f}}} (u_\tau + M^{-1}[V^{\mathrm{T}} + B^{\mathrm{T}}P(\tau)]z_\tau)^{\mathrm{T}}$$

$$\times M(u_\tau + M^{-1}[V^{\mathrm{T}} + B^{\mathrm{T}}P(\tau)]z_\tau)\,d\tau$$

$$+ \frac{1}{2}\int_0^{\hat{T}_{\mathrm{f}}} dv_\tau^{\mathrm{T}}\,C_\tau^{\mathrm{T}}P(\tau)z_\tau + \frac{1}{2}\int_0^{\hat{T}_{\mathrm{f}}} z_\tau^{\mathrm{T}}P(\tau)C\,dv_\tau$$

$$+ \frac{1}{2} \int_0^{\hat{T}_f} \operatorname{tr} P(\tau) C R_0 C^T \, d\tau, \tag{B.3}$$

where $\operatorname{tr} P(\tau)CR_0C^T$ denotes the trace operation on $P(\tau)CR_0C^T$, i.e., the sum of the diagonal terms of $P(\tau)CR_0C^T$.

Since z_t is a solution of a stochastic differential equation it does not have a time derivative. Ito calculus will, therefore, be used to prove the Lemma. We shall need the Ito differentiation rule which is defined below.

DEFINITION. (Ito Differentiation Rule [2]). Let the n-vector x satisfy the stochastic differential equation

$$dx = f(x, t)\, dt + \sigma(x, t)\, dw,$$

where $\{w_t, t \in T\}$ is a Wiener process with incremental covariance $I\, dt$. Let the function $y(x, t)$ be continuously differentiable in t and twice continuously differentiable in x. Then y satisfies the following stochastic differential equation:

$$\begin{aligned} dy &= \frac{\partial y}{\partial t} dt + \sum_{i=1}^{n} \frac{\partial y}{\partial x_i} dx_i + \frac{1}{2} \sum_{i,j,k=1}^{n} \frac{\partial^2 y}{\partial x_i\, \partial x_j} \sigma_{ik}\sigma_{jk}\, dt \\ &= \left(\frac{\partial y}{\partial t} + \sum_{i=1}^{n} \frac{\partial y}{\partial x_i} f_i + \frac{1}{2} \sum_{i,j,k=1}^{n} \frac{\partial^2 y}{\partial x_i\, \partial x_j} \sigma_{ik}\sigma_{jk}\right) dt \\ &\quad + \sum_{i=1}^{n} \frac{\partial y}{\partial x_i} (\sigma dw)_i . \end{aligned} \tag{B.4}$$

<u>Proof</u>. We can write, using $P(T_f) = K$,

$$\frac{1}{2} z^T(T_f) K z(T_f) = \frac{1}{2}(z^T(0)P(0)z(0) + \int_0^{T_f} d(z_\tau^T P(\tau) z_\tau)). \tag{B.5}$$

Using the Ito differentiation rule to calculate $d(z^{\mathrm{T}}Pz)$ we have

$$d(z_t^{\mathrm{T}}P(t)z_t) = z_t^{\mathrm{T}} \frac{dP(t)}{dt} z_t\, dt + dz_t^{\mathrm{T}}P(t)z_t + z_t^{\mathrm{T}}P(t)\, dz_t + \mathrm{tr}\, P(t)CR_0C^{\mathrm{T}}\, dt. \tag{B.6}$$

Substituting $\frac{dP(t)}{dt}$ by Eq. (B.8) and dz_t in the second and third terms by Eq. (B.1) we obtain

$$\begin{aligned} d(z_t^{\mathrm{T}}P(t)z_t) = & -(z_t^{\mathrm{T}}Lz_t + u_t^{\mathrm{T}}Mu_t + z_t^{\mathrm{T}}Vu_t + u_t^{\mathrm{T}}V^{\mathrm{T}}z_t)\, dt \\ & + dt(u_t^{\mathrm{T}}Mu_t + z_t^{\mathrm{T}}(V + P(t)B)u_t + u_t^{\mathrm{T}}(V^{\mathrm{T}} + B^{\mathrm{T}}P(t))z_t \\ & + z_t^{\mathrm{T}}(P(t)BM^{-1}V^{\mathrm{T}} + VM^{-1}B^{\mathrm{T}}P(t) + P(t)BM^{-1}B^{\mathrm{T}}P(t) \\ & + VM^{-1}V^{\mathrm{T}})z_t) + dv_t^{\mathrm{T}}\, C^{\mathrm{T}}P(t)z_t + z_t^{\mathrm{T}}P(t)C\, dv_t \\ & + \mathrm{tr}\, P(t)CR_0C^{\mathrm{T}}\, dt. \end{aligned} \tag{B.7}$$

Performing the operation of completion of squares we obtain

$$\begin{aligned} d(z_t^{\mathrm{T}}P(t)z_t) = & (u_t + M^{-1}(V^{\mathrm{T}} + B^{\mathrm{T}}P(t))z_t)^{\mathrm{T}}\, M(u_t + M^{-1}(V^{\mathrm{T}} \\ & + B^{\mathrm{T}}P(t))z_t)\, dt + dv_t^{\mathrm{T}}\, C^{\mathrm{T}}P(t)z_t \\ & + z_t^{\mathrm{T}}P(t)C\, dv_t + \mathrm{tr}\, P(t)CR_0C^{\mathrm{T}}\, dt \\ & - (z_t^{\mathrm{T}}Lz_t + u_t^{\mathrm{T}}Mu_t + z_t^{\mathrm{T}}Vu_t + u_t^{\mathrm{T}}V^{\mathrm{T}}z_t)\, dt. \end{aligned} \tag{B.8}$$

Substituting (B.8) into (B.5) yields (B.3).

APPENDIX C. STABILITY OF THE $\hat{u}^{s}_{II}$ -SYSTEM

The suboptimal zero input stochastic suspension system Case II:

$$\dot{\hat{x}}_t = A^{s}_{sII}\hat{x}$$

$$A^{s}_{sII} = [I - B_s M^{-1} B_s^{T} S] A_s - B_s M^{-1} B_s^{T} \Pi_{11}$$

is uniformly asymptotically stable.

Proof. The optimal closed-loop system for complete state feedback was given by Eq. (38) to be

$$dz_t = A^{0} z_t \, dt + C \, dv_t$$

where $A^{0} = A - BM^{-1}[B^{T}P(t) + V^{T}]$. We use the following theorem.

THEOREM A. The optimal closed-loop system (38) with zero input is uniformly asymptotically stable.

Proof. Twice adding and subtracting the term $P(t)BM^{-1}B^{T}P(t)$ in Eq. (37) we obtain

$$\frac{dP(t)}{dt} = -P(t)A^{0} - A^{0T}P(t) + VM^{-1}V^{T} - L - P(t)BM^{-1}B^{T}P(t).$$

Since $\lim_{T_f \to \infty} P(t) = \Pi$ exists and is positive definite we have

$$VM^{-1}V^{T} - L - \Pi BM^{-1}B^{T}\Pi = -Q^{0},$$

$$\bar{0} = \Pi A^{0} + A^{0T}\Pi + Q^{0}. \qquad \text{(C.1)}$$

Equation (C.1) is the defining equation for the Liapunov function $V(z) = z^{T}\Pi z$. Hence A^{0} is stable if and only if Π exists and is positive definite, and Q^{0} is positive definite [5]. It remains only to show $Q^{0} > 0$.

Define

$$L_{V} = L - VM^{-1}V^{T}.$$

Then L_{V} is positive definite if L is positive definite. This follows from the definition of L, M, and V as given in Eq. (32). We have

$$L - VM^{-1}V^{T} = [A_{s} : C_{s}]^{T}(S - SB_{s}[B_{s}^{T}SB_{s} + R]^{-1}B_{s}^{T}S)[A_{s} : C_{s}] + Q,$$

$$S > SB_{s}[B_{s}^{T}SB_{s} + R]^{-1}B_{s}^{T}S \quad \text{since } R > 0,$$

$$Q^{0} = L_{V} + \Pi BM^{-1}B^{T}\Pi.$$

Since $L_{V} > 0$ and $\Pi BM^{-1}B^{T}\Pi$ is nonnegative definite we see that Q^{0} is positive definite.

We obtain the following suboptimal stochastic vehicle equation by substituting $\hat{u}_{II}^{s}$ for $\hat{u}$ in Eq. (17):

$$\dot{\hat{x}}_{t} = A_{s}\hat{x}_{t} + B_{s}\hat{u}_{II}^{s}(t) + C_{s}\hat{g}_{t}. \tag{C.2}$$

Substituting $\hat{u}_{II}^{s}(t) = -M^{-1}(B^{T}\tilde{\Pi}\tilde{D}D + V^{T})z_{t}$ into Eq. (C.2) we obtain with $\tilde{D}D$ as defined in Eq. (71)

$$\dot{\hat{x}}_{t} = ([I - B_{s}M^{-1}B_{s}^{T}S]A_{s} - B_{s}M^{-1}B_{s}^{T}\Pi_{11})\hat{x}_{t}$$

$$+ ([I - B_{s}M^{-1}B_{s}^{T}S]C_{s} - B_{s}M^{-1}B_{s}^{T}\Pi_{11}D_{3})\hat{g}_{t}.$$

with initial conditions as in Eq. (17).

Partitioning A^0 we obtain for $t \to \infty$

$$A^0 = \begin{bmatrix} A_s - B_s M^{-1}(B_s^T \Pi_{11} + B_s^T S A_s) & C_s - B_s M^{-1}(B_s^T \Pi_{12} + B_s^T S C_s) \\ 0 & F_g \end{bmatrix}$$

By Theorem A $dz_t = A^0 z_t \, dt$ is uniformly asymptotically stable.

Since $A_{11}^0 = A_{sII}^s$ and the partition Π_{11} of $\tilde{\Pi}$ is positive definite the conclusion of the theorem follows.

This theorem is significant because it alleviates the need for a stability check of the suboptimal stochastic system if $\hat{u}_{II}^s$ has been chosen as the suboptimal control. Moreover, it is seen that the eigenvalues of the suboptimal system Case II are identical to the eigenvalues of the optimal system with unconstrained state feedback.

ECONOMIC SYSTEMS

MICHAEL D. INTRILIGATOR

Economics Department
AMSAR/RDT, U.S. Army Armaments Command
Rock Island, Illinois

I. ECONOMICS

Economics is concerned with both a process and a set of

institutions. The process, called the economizing process, is a study of the allocation of scarce resources among competing ends, to be discussed in Section II. The institutions are those of the economy, to be discussed in Section III. Later sections give examples of the application of the economizing process to specific institutions of the economy and other problem areas.

II. STATIC AND DYNAMIC PROBLEMS OF ECONOMIZING

The economizing process of allocating scarce resources among competing ends can be represented formally as a mathematical problem of optimization. (For a detailed discussion of this approach to economic analysis, including the development of the mathematical problem and its applications to economics see, Intriligator, [1].) Any particular allocation can be summarized by a set of variables, called instruments. These are the variables controlled by the decision-maker facing the problem of economizing. The fact that resources are "scarce" can be represented by the fact that the variables controlled by the decision maker cannot be chosen freely but rather must be chosen within certain limits, represented by the opportunity set. The competing ends of the problem and the relative importance of these ends can be summarized by a relation indicating the value to the decision maker of alternative allocations. This relation is called an objective function. The economizing process can thus be represented as the mathematical problem

$$\max_{\underset{\sim}{x}} F(\underset{\sim}{x}) \quad \text{subject to} \quad \underset{\sim}{x} \in X, \tag{1}$$

where $\underset{\sim}{x}$ is a vector of instruments, X the opportunity set of feasible instruments, and $F(\underset{\sim}{x})$ the objective function. The economizing process can thus be represented as the mathematical problem of choosing instruments within the opportunity set so as to maximize the objective function--the problem of mathematical optimization.

Two particular formulations of the mathematical optimization problem (1) have been of fundamental concern to economists, mathematicians, and engineers.

The first formulation of (1) is the static problem of mathematical programming, in which $\underset{\sim}{x}$ is a column vector of real numbers, X a subset of Euclidean space, and $F(\underset{\sim}{x})$ a real-valued function defined on X. The problem is then one of choosing values for the instruments within a set of possible values so as to maximize the objective function. It includes, as special cases, problems of classical programming, linear programming, and nonlinear programming.

The problem of nonlinear programming is of the form

$$\max_{\underset{\sim}{x}} \; F(\underset{\sim}{x}) \qquad \text{subject to} \qquad \underset{\sim}{g}(\underset{\sim}{x}) \leq \underset{\sim}{b},\; \underset{\sim}{x} \geq \underset{\sim}{0}\;, \tag{2}$$

where $\underset{\sim}{g}(\underset{\sim}{x})$ represents a set of constraint functions and $\underset{\sim}{b}$ is a set of constraint constants. According to the Kuhn-Tucker theorem, defining the Lagrangian function as

$$L(\underset{\sim}{x}, \underset{\sim}{y}) \;=\; F(\underset{\sim}{x}) + \underset{\sim}{y}\cdot(\underset{\sim}{b} - \underset{\sim}{g}(\underset{\sim}{x})), \tag{3}$$

where the row vector $\underset{\sim}{y}$ consists of Lagrange multipliers, necessary conditions for a particular set of instruments $\underset{\sim}{x}$ to solve (2) are

$$\begin{aligned} &\frac{\partial L}{\partial \underset{\sim}{x}} \leq \underset{\sim}{0}\,, && \frac{\partial L}{\partial \underset{\sim}{y}} \geq \underset{\sim}{0}\,, \\ &\frac{\partial L}{\partial \underset{\sim}{x}} \cdot \underset{\sim}{x} = 0\,, && \underset{\sim}{y} \cdot \frac{\partial L}{\partial \underset{\sim}{y}} = 0\,, \\ &\underset{\sim}{x} \geq \underset{\sim}{0}\,, && \underset{\sim}{y} \geq \underset{\sim}{0}\,. \end{aligned} \tag{4}$$

The solutions for the Lagrange multipliers can then be interpreted as the sensitivities of the maximized value of the objective function to changes in the constraint constants. Thus

$$\underset{\sim}{y} = \frac{\partial F^*}{\partial \underset{\sim}{b}}. \tag{5}$$

These Lagrange multipliers in many problems of economic allocation have the interpretation of shadow prices, that is, prices that guide an allocation but which generally are not prices observed on actual markets. If the objective function has the dimension of a value, that is, a price times a quantity, such as income, cost, revenue, and profit, and if the constraint constant has the dimension of a quantity, such as output, labor, or capital, then the Lagrange multiplier has the dimension of a price--a shadow price. This price interpretation is an important one in many problems of economizing since it indicates the "value"--in terms of the ojective function--of relaxing the constraint.

The second formulation of (1) is the dynamic problem of mathematical control, defined over time t ranging from t_0, the initial time, to t_1 the terminal time. The instruments are the trajectories over time $\{\underset{\sim}{u}(t)\}$, defined over the interval $t_0 \leq t \leq t_1$, of a set of control variables. The opportunity set U is the control set, a subset of function space, the space of all such trajectories over time. The objective functional $J\{\underset{\sim}{u}(t)\}$ is then a mapping from U into the real line. The problem is usually formulated as

$$\max_{\{\underset{\sim}{u}(t)\}} J\{\underset{\sim}{u}(t)\} = \int_{t_0}^{t_1} I(\underset{\sim}{x}, \underset{\sim}{u}, t)\, dt + F(\underset{\sim}{x}_1, t_1), \tag{6}$$

$$\dot{\underset{\sim}{x}} = \underset{\sim}{f}(\underset{\sim}{x}, \underset{\sim}{u}, t),$$

$$t_0,\ \underset{\sim}{x}(t_0) = \underset{\sim}{x}_0, \qquad \text{given}$$

$$t_1 \text{ given}, \quad \underset{\sim}{x}(t_1) = \underset{\sim}{x}_1,$$

$$\{\underset{\sim}{u}(t)\} \in U$$

where $\underset{\sim}{x}(t)$ defines the state of the system at time t; $\underset{\sim}{f}(\underset{\sim}{x}, \underset{\sim}{u}, t)$ represents the equations of motion, determining the time rates of change of the state variables; $\underset{\sim}{x}_0$ and $\underset{\sim}{x}_1$ refer to the initial and terminal states, respectively; $I(\underset{\sim}{x}, \underset{\sim}{u}, t)$ is the intermediate objective, giving the contribution to the objective functional over the interval between t and $t + dt$; and $F(\underset{\sim}{x}_1, t_1)$ is the final objective, giving the contribution to the objective functional at terminal time. The problem is then one of choosing trajectories over time for the control variables within the set of admissible trajectories so as to influence the objective functional both directly and indirectly--the latter by influencing the trajectories for the state variables--in such a way that the objective functional is maximized. The solution is an optimal control. There are several approaches to obtaining such an optimal control, including the calculus of variations, dynamic programming, and the maximum principle.

According to the maximum principle, defining the Hamiltonian function as

$$H(\underset{\sim}{x}, \underset{\sim}{u}, \underset{\sim}{y}, t) \equiv I(\underset{\sim}{x}, \underset{\sim}{u}, t) + \underset{\sim}{y} \cdot \underset{\sim}{f}(\underset{\sim}{x}, \underset{\sim}{u}, t), \tag{7}$$

where $\underset{\sim}{y}$ consists of costate variables, the dynamic equivalents of the Lagrange multipliers of static problems of mathematical programming, necessary conditions for a particular control trajectory $\{\underset{\sim}{u}(t)\}$ to solve (7) are

$$\max_{\underset{\sim}{u}} H(\underset{\sim}{x}, \underset{\sim}{u}, \underset{\sim}{y}, t) \text{ for all } t, \quad t_0 \leq t \leq t_1, \tag{8}$$

$$\dot{\underset{\sim}{x}} = \frac{\partial H}{\partial \underset{\sim}{y}}, \qquad \underset{\sim}{x}(t_0) = \underset{\sim}{x}_0,$$

$$\dot{\underset{\sim}{y}} = -\frac{\partial H}{\partial \underset{\sim}{x}}, \qquad \underset{\sim}{y}(t_1) = \frac{\partial F}{\partial \underset{\sim}{x}_1}.$$

Thus at each point in time the control vector $\underset{\sim}{u}$ maximizes the Hamiltonian of (7). The resulting trajectory $\{\underset{\sim}{u}^*(t)\}$ is the optimal control. The maximization called for in (8) is an ordinary maximization of a function--the Hamiltonian function--but this function depends on $\underset{\sim}{x}$ and $\underset{\sim}{y}$ as well as $\underset{\sim}{u}$ (and t). The behavior of $\underset{\sim}{x}$ and $\underset{\sim}{y}$ is determined by the coupled system of differential equations in (8), called the canonical equations, together with appropriate boundary conditions. The differential equations and boundary conditions for $\underset{\sim}{x}$ are the same as in (6), but those for $\underset{\sim}{y}$ provide new information on the behavior of the costate variables. The canonical equations are systems of differential equations with separated boundary conditions, so their solution entails solving a two-point boundary value problem. If this problem can be solved, however, the problem reduces from one of control to one of programming--that of maximizing the Hamiltonian at each instant of time.

An important reason why economists tend to utilize the maximum principle approach is the interpretation given the costate variables. Just as in the static case, in many problems of economizing they have the interpretation of shadow prices, giving the "value"--in terms of the objective functional--of increasing the corresponding state variable at that time. This interpretation is clear from the boundary condition for $\underset{\sim}{y}$ at terminal time, but it holds more generally at any time. At any time, then, the state variables play a role that is analogous to the constraint constants of the nonlinear programming problem. Over time the trajectories for the costate variables give the evolution of the shadow prices, which guide the optimal allocation at each instant by affecting the value of the Hamiltonian.

III. THE ECONOMY

The economy consists of various agents and organizations which interact so as to determine economic phenomena, such as production, employment, income, and (market) prices. An important distinction can be drawn between microeconomic and macroeconomic agents and organizations. Microeconomic agents and organizations are individual decision makers solving economizing problems. Macroeconomic agents and organizations, by contrast, solve economizing problems involving aggregates of individual economic variables. Section IV gives examples of microeconomic economizing problems, while Section V gives an example of a macroeconomic economizing problem. The next two sections present applications of the economizing problem to areas not traditionally identified with economics but nevertheless entailing problems of allocating scarce resources.

IV. THE HOUSEHOLD AND THE FIRM

The two most important microeconomic institutions are the household and the firm, and each solves an economizing problem [1, 2].

The economizing problem faced by the household (or consumer) is that of maximizing utility subject to a budget constraint. Letting $\underset{\sim}{x} = (x_1 x_2 \cdots x_n)'$ be the (column) vector of n goods and services purchased by the household, $U(\underset{\sim}{x})$ be the utility function of the household, $\underset{\sim}{p} = (p_1 p_2 \cdots p_n)$ be the (row) vector of prices of the goods and services purchased, and I be the household income, the economizing problem of the household can be expressed as

$$\max_{\underset{\sim}{x}} \; U(\underset{\sim}{x}) \quad \text{subject to} \quad \underset{\sim}{p}\underset{\sim}{x} \leq I. \tag{9}$$

The problem is one of maximizing utility by choice of purchases subject to the budget constraint that expenditure not exceed income. This formulation is the same form as the nonlinear programming problem (2), and the solution in (4) implies that for all goods purchased

$$\frac{\partial U}{\partial x_j} = y p_j \,. \tag{10}$$

Thus among the goods purchased the ratios of marginal utility (the partial derivative) to price are all equal. This common ratio is the Lagrange multiplier y, where, using (5),

$$y = \frac{\partial U^*}{\partial I}, \tag{11}$$

that is, the Lagrange multiplier is the marginal utility of income. The conditions in (10) and the equality budget constraint

$$\underset{\sim}{p}\underset{\sim}{x} = \sum_{j=1}^{n} p_j x_j = I \tag{12}$$

jointly determine the quantities demanded and the marginal utility of income as functions of all prices $p_1, p_2, \ldots, p_n$, and income I.

The second microeconomic institution, the firm, faces the economizing problem of maximizing profit subject to a technological constraint. Letting $\underset{\sim}{x} = (x_1 x_2 \cdots x_n)'$ be the (column) vector of n inputs (factors of production) which can be purchased by the firm; q be the output of the firm; $f(\underset{\sim}{x})$ be the production function, giving output as a function of all inputs; p be the price of output; and $\underset{\sim}{w} = (w_1 w_2 \cdots w_n)$ be the (row) vector of wages of the inputs, the economizing problem of the firm can be expressed as

$$\max_{q,\underset{\sim}{x}} \; pq - \underset{\sim}{w}\underset{\sim}{x} \quad \text{subject to} \quad q \leq f(\underset{\sim}{x}). \tag{13}$$

This problem is one of maximizing profits, equal to revenue less cost, by choice of output and inputs, subject to the constraint of the technology embodied in the production function. Again the formulation is as in (2), and the solution in (4) implies that for all inputs purchased

$$y \frac{\partial f}{\partial x_j} = w_j \; . \tag{14}$$

Thus among the inputs purchased the ratios of marginal physical product (the partial derivative) to wage are all equal, the common ratio being the reciprocal of the Lagrange multiplier y. In this case the Lagrange multiplier is the same as output price

$$y = p, \tag{15}$$

so (14) states that among inputs purchased the marginal physical product is relative price, that is, the ratio of the wage to output price w_j/p. These conditions determine the quantities demanded of all inputs as functions of all relative prices. These demand functions for inputs imply a supply function for output, obtained by combining the demand functions with the production function.

For both the household and the firm it is possible to vary the parameters, particularly the prices and wages, in order to determine the implied effects on the quantities demanded. This technique, called <u>comparative statics</u>, restricts the values of certain partial derivatives of the quantities demanded. These theoretical restrictions are used both in analytical studies and in empirical investigations.

V. OPTIMAL ECONOMIC GROWTH

Macroeconomics is concerned with aggregate economic phenomena such as total income, total consumption, and the total capital stock or the corresponding per capita variables in order to focus on a representative individual. An important problem facing the aggregate economy is that of how fast to grow, given that growth and future consumption depend on capital formation, but capital formation entails less present consumption. (The original reference, which was years ahead of its time, was by Ramsey [3]; see also Shell [4], Arrow [5], and Shell [6].) One extreme policy would be to "tighten the belt", holding current consumption down to minimal levels in order to ensure rapid capital formation and high potential consumption in the future. The other extreme policy would be to "splurge", consuming large amounts today, without regard for the future. The problem of optimal economic growth is that of finding an optimal consumption policy over time.

The simplest optimal economic growth problem, that for a one sector closed economy with a single homogeneous good is

$$\max_{\{c(t)\}} W = \int_{t_0}^{\infty} e^{-\delta(t-t_0)} U(c(t))\, dt, \tag{16}$$

$$\dot{k} = f(k) - \lambda k - c, \qquad k(t_0) = k_0,$$

$$0 \leq c \leq f(k), \qquad c(t) \text{ piecewise continuous.}$$

Here W is the social welfare functional, to be maximized by choice of a trajectory $\{c(t)\}$ for consumption per worker. The functional is defined as the sum (integral) of utilities derived from consumption per capita $U(c)$, discounted at some positive rate δ over a time horizon from the initial time t_0 through all time. The state variable is capital per worker k measured as the stock of capital with which each worker produces output.

According to the equation of motion the rate of change over time of k is given as the output per worker, defined by the production function $f(k)$ less the constant λ times k and less consumption per worker. The constant λ is the sum of the depreciation rate of capital and the rate of growth of population, since capital per worker is eroded due to both factors. Both are assumed constant here. The economy starts from a given level of capital per worker k_0 and consumption per worker at any time can range between a minimum of zero ("tighten the belt") and a maximum, for a closed economy, of total output per worker ("splurge"). Furthermore, the function $c(t)$ should be piecewise continuous, i.e., continuous other than a finite number of finite jumps. These conditions on $c(t)$ define the control set for the problem. The utility function $U(c)$ and the production function $f(k)$ are both assumed to be monotonically increasing, strictly concave, and twice differentiable functions, and these two functions, plus the four positive parameters δ, λ, t_0, and k_0, define the problem of optimal economic growth.

Using the maximum principle approach, the Hamiltonian can be defined as

$$H = e^{-\delta(t-t_0)}\{U(c) + q[f(k) - \lambda k - c]\}, \tag{17}$$

where q is the costate variable.[†] The costate here has the interpretation of the imputed value (shadow price) of additional capital per worker, measured in terms of utility. The Hamiltonian then has the interpretation of the discounted value of output per worker, composed of discounted utility derived from consumption and discounted *imputed* utility derived from capital formation. The costate variable measures the utility obtained from current capital formation due to its increasing

[†]In terms of (7), the variable deflated for discounting $qe^{-\delta(t-t_0)}$ is the costate y. It is convenient, however, to refer to the undeflated variable q as the costate.

future consumption.

Maximizing the Hamiltonian by choice of the control variable c requires that

$$q = U'(c) = \frac{dU(c)}{dc}, \tag{18}$$

so the shadow price of capital accumulation along the optimal path is optimally the marginal utility derived from added consumption per worker. Combining (18) and the canonical equations of this problem leads to the conclusion that if the paths $\{c^*(t)\}$ and $\{k^*(t)\}$ are optimal, they must satisfy the differential equations†

$$\dot{c} = \frac{1}{\sigma(c)} [f'(c) - (\lambda+\delta)]c$$

$$\dot{k} = f(k) - \lambda k - c. \tag{19}$$

The nature of the solution is indicated in Fig. 1, where part (a) shows $f(k)$, output per worker, and λk as functions of k, capital per worker. The curve $f(k) - \lambda k$ in part (b), obtained as the difference of the curve and line in part (a), gives $\dot{k} + c$ according to the equation of motion. It reaches a maximum at $\hat{k}$ where $f'(k) = \lambda$, called the golden rule, yielding the maximum sustainable level of consumption per worker at $\hat{c} = f(\hat{k}) - \lambda\hat{k}$.

The point marked with an asterisk * is the balanced growth

---

† Here $\sigma(c)$ is the elasticity of marginal utility, defined as

$$\sigma(c) = -c\,\frac{U''(c)}{U'(c)}.$$

It is positive for all positive c, since $U''(c) < 0$ and $U'(c) > 0$.

<u>equilibrium</u>, defined as the (k^*, c^*) pair satisfying

$$f'(k^*) = \lambda + \delta, \qquad c^* = f(k^*) - \lambda k^*. \tag{20}$$

From (19) at the balanced growth equilibrium $\dot{c}$ and $\dot{k}$ are

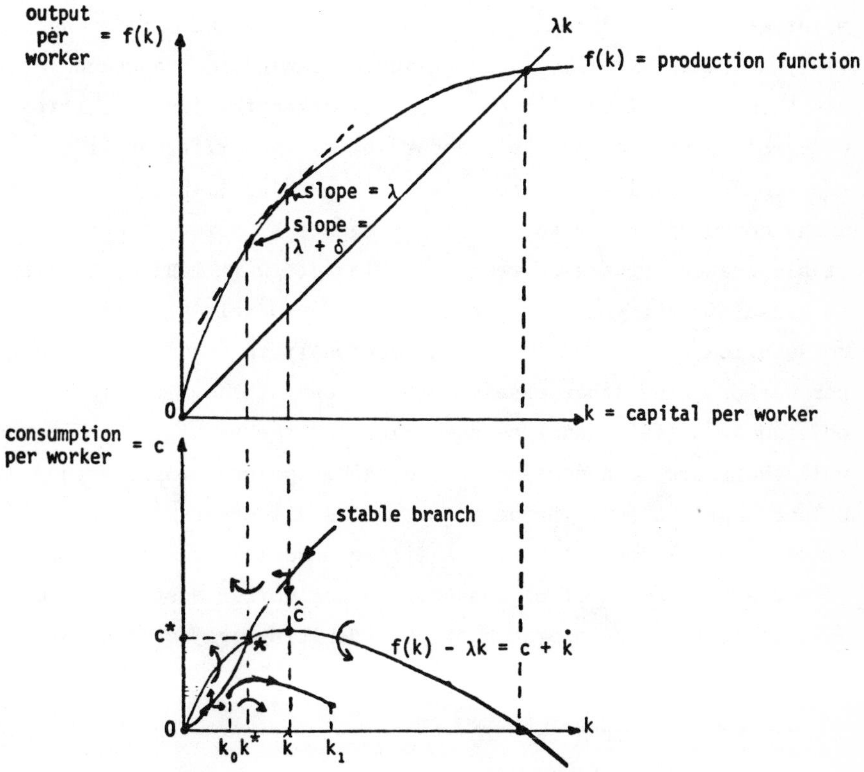

Fig. 1. Optimal Economic Growth

both zero, so at this point both capital per worker and consumption per worker are optimally constant. Once this point is reached the system will optimally remain at it. This point is called the modified golden rule since it modifies the golden rule to allow for the discount rate δ. The modification is indicated in Fig. 1a.--at the golden rule level of capital per worker the slope of $f(k)$ is λ while at the modified golden rule it is $\lambda + \delta$.

The differential equations in (19) imply, for optimality, that capital per worker and consumption worker should asymptotically approach the balanced growth equilibrium levels of k^* and c^*, respectively. They also imply optimal directions of motion, indicated by arrows. Finally, they imply that the equilibrium point at (k^*, c^*) is a saddle point of the paths satisfying (19), as shown in Fig. 1b. The stable branch shows at each possible level of initial capital per worker the level of initial consumption per worker that, according to the differential equations, will lead to the balanced growth equilibrium.[†] "Too much" consumption per worker, i.e., choice of points above the stable branch for the control variable c, will lead eventually to "insufficient" capital per worker. Conversely "too little" consumption per worker will lead eventually to "excessive" capital per worker. In either case eventually one of the differential equations in (19) cannot be satisfied and the resulting trajectory will therefore be nonoptimal. The saddle point nature of the balanced growth equilibrium means that small errors in the choice of the initial control variable will eventually be magnified. Thus the initial level of consumption per worker must be chosen exactly on the stable branch or the resulting trajectories will be nonoptimal.

[†] Note that the differential equations in (19) and the given initial capital per worker $k(t_0)$ completely define both $\{c(t)\}$ and $\{k(t)\}$ once $c(t_0)$ is specified.

Consider now the finite time problem, where the upper limit of the integral in (16) is finite, specified as the terminal time t_1. It is then necessary to specify, in addition to initial capital per worker, a minimum level for terminal capital per worker, of the form

$$k(t_1) \geq k_1. \tag{21}$$

This minimum level k_1 indicates what the future generations starting at t_1 can be guaranteed as their initial level of capital per worker. The solution to this finite time problem can also be illustrated in Fig. 1, as in the path from k_0 to k_1. In this case the economy starts from a low level of capital per worker and ultimately must attain a relatively high level of capital per worker. To grow optimally still entails the same differential equations of (19) so the directions of motion, indicated by the arrows, are the same as before. In this case, however, less initial consumption per worker than that indicated by the stable branch is optimal in order to build up capital per worker to the stipulated minimum terminal level. Note, however, that the optimal path initially moves toward the balanced growth equilibrium, as in the infinite time case, and eventually moves away from it in order to satisfy the terminal condition (21). This movement toward the balanced growth equilibrium is referred to as the turnpike property of the optimal path. The optimal path does not move directly from the initial capital per worker to the terminal capital per worker; rather it arcs toward the balanced growth equilibrium and then away from it in order to satisfy the terminal requirement. The "turnpike" name is an apt one: instead of driving one's automobile directly from one point to another it is usually optimal to go somewhat out of the way to reach a turnpike, move along this optimal path for a while, and then finally move away from the turnpike in order to reach the final destination. Optimal capital accumulation is exactly like this, where the turnpike is the balanced growth equilibrium. Note that the longer the time in the problem (t_1-t_0) and the closer the terminal capital per worker to k^*, the closer will the optimal path be to the optimal path in the infinite time problem, i.e., movement along the stable path to the balanced growth equilibrium.

There are many variants and extensions of the basic problem but they all involve solutions that are qualitatively similar to that described here. (For discussions of variants of the basic problem, entailing a two-sector economy, heterogeneous capital goods, foreign trade, etc., see the work of Shell [4,6] and Intriligator [1].) First, the costate variable(s) are shadow prices. Second, there exists a balanced growth equilibrium, but it is a saddle point for the optimal system of differential equations. Third, the finite-time problem solutions exhibit the turnpike property of arcing initially toward the balanced growth equilibrium and eventually away from it in order to satisfy the requirements at terminal time.

VI. SCIENCE POLICY

Another problem for economic analysis is that of science policy. While no one would doubt that economic growth, the subject of the last section, is a problem of economics, some may express surprise that science policy is also one of economics. The definition of economics as the science of allocating scarce resources, however, does indeed suggest that science policy is a proper area of inquiry for economists; scientists are a scarce resource, and their optimal allocation may be indicated via the tools of economic analysis.

One important issue of science policy is the allocation of new scientists between teaching and research [7,8]. In particular, the "feedback" of science doctorates into higher education raises important issues of policy. There may be a danger of seriously weakening the educational process or, alternatively, of "starving" nonacademic research endeavors by an improper allocation between scientific careers. One approach to analyzing these issues formulates the problem as one of solving the following control problem [7]:

$$\max_{\{\beta(t)\}} W = F(E(t_1), R(t_1)), \tag{22}$$

$$\dot{E} = \beta g E(t) - \delta E(t), \qquad E(t_0) = E_0,$$

$$\dot{R} = (1-\beta) g E(t) - \delta R(t), \qquad R(t_0) = R_0,$$

$$\beta_0 \leq \beta \leq \beta_1 , \qquad \beta(t) \text{ piecewise continuous.}$$

In this formulation $E(t)$ and $R(t)$ are the state variables, representing, respectively, teaching (educator) scientists and research scientists at time t.† In the equations of motion, g is the number of scientists annually produced, on the average, by one teaching scientist, and δ the rate of exit, via death, retirement, or transfer, from science.‡ The differential equations determine the rates of change of the numbers of teaching and research scientists as new scientists allocated to each less the losses due to death, retirement, or transfer. Initial numbers of teaching and research scientists are given as E_0 and R_0, respectively. The variable β is the control variable, representing the proportion of new

---

† Of course some scientists are involved in both teaching and research, such as scientists at leading universities. Thus E and R should be measured in terms of full time equivalents.

‡ It has been assumed that once scientists become teaching or research scientists they do not switch to the other career.

scientists becoming teachers, the remaining proportion $(1 - \beta)$ representing the proportion becoming researchers.

The problem is one of choosing a trajectory for this allocation proportion $\{\beta(t)\}$ where, at any time, β must lie between certain minimum and maximum values, β_0 and β_1, respectively, and, over time, $\beta(t)$ must be piecewise continuous. The extreme values β_0 and β_1 indicate the limits of science policy in affecting initial career choices, by means of grants, provisions of government contracts, etc.

The objective function in (22) depends on the numbers of teaching and research scientists at a time horizon t_1, where $t_1 > t_0$. This is a broad objective function that includes several possible objectives of science policy as special cases. One such case is that of maximizing the value of scientific effort at a given future date, given the value of teachers relative to researchers as of this date. Another is to minimize the time required to attain given terminal numbers of teaching and research scientists.

Solving the problem in (22) by means of the maximum principle leads to the introduction of costate variables, which, as before, have the interpretation of shadow prices. Here the costate variables represent the marginal social benefits of scientists as teachers and the marginal social benefits of scientists as researchers, respectively. The solution to the problem requires that the maximum number of new scientists be allocated to teaching $(\beta = \beta_1)$ if the marginal social benefit of teachers exceeds that of researchers and, conversely, the maximum allocated to research $(\beta = \beta_0)$ if the marginal social benefit of researchers is higher. Combining this result with the solutions of the differential equations for the costate variables, from the canonical equations, results in the optimal time path for the allocation proportion

$$\beta^*(t) = \begin{cases} \max = \beta_1 & \text{if} \quad t_0 \le t \le t^* \\ \min = \beta_0 & \quad\quad t^* < t \le t_1 \end{cases} . \tag{23}$$

This solution entails allocating the maximum proportion of new scientists to teaching during an initial period and then switching to allocate the minimum proportion to teaching (and the maximum proportion to research) during a terminal period. The switching time t^*, when the allocation proportion switches, is determined by the parameters of the problem, g, δ, t_1, β_0, β_1, and the parameters determining the objective function.

The solution entails initially allocating the maximum proportion of new scientists to teaching and then switching so as to allocate the maximum to research. Such an optimal path illustrates the turnpike property of initial movement toward the turnpike, which here would allocate all new scientists to teaching in order to increase the total number of scientists, followed by terminal movement toward the desired goal, as in the case of optimal economic growth. Such a path is preferable to one that moves directly to the target.

To give a numerical example, assume the problem is one of minimizing the time required to move from 100 teachers and 80 researchers to 200 teachers and 240 researchers. Assume that each teaching scientist produces 0.14 new scientists per year (e.g., on the average, seven doctorate scientist educators produce one new science doctor each year), scientists exit their profession at the rate of 2% per year, and, through appropriate policies, the percent of new scientists choosing a teaching career can range from 10 to 60%.† The solution to this problem allocates the maximum (60) percent of new scientists to teaching

†The estimate of 0.14 for g was obtained by Bolt and co-workers [8].

in the first 11.2 years and then switches so as to allocate the minimum (10) percent to teaching in the remaining 4.3 years, attaining the desired terminal numbers of teachers and researchers in 15.5 years. Any other allocation policy would entail a longer period to reach the desired target numbers of teachers and researchers.

The switching nature of the analytic results is a point that merits reflection. The desired objectives are best achieved by allocating first a maximum and then a minimum proportion of new scientists to teaching. This conclusion is hardly surprising to economists familiar with unbalanced growth and turnpike theorems. It might, however, appear rather novel to policy makers and observers of science policy who have often considered balance and gradualism important components of science policy, but the notion of pronounced shifts in science policy actually has some intuitive appeal. Science often does advance in an unbalanced pattern, making rapid strides first in some sectors and then in others. Indeed it is often the case that science is most productive when it advances in such a way. A high degree of flexibility and an ability to "shift gears" quickly may, in fact, be the hallmarks of a successful science policy.

VII. MILITARY STRATEGY

Another application of the economizing problem is military strategy. Again this is an area not traditionally identified as one of economics but nevertheless one in which the problem is fundamentally that of allocating scarce resources among competing ends. Thus, in fighting a missile war, a given number of missiles available at the outset of the war must be allocated both over time and between alternative targets.

One approach to military strategy analyzes optimal strategy in terms of the following control problem [9-11]:

$$\max_{\{\alpha(t),\alpha'(t)\}} P_A(M_A(t_1),\ M_B(t_1),\ C_A(t_1),\ C_B(t_1)), \tag{24}$$

$$\dot{M}_A = -\alpha M_A - \beta M_B \beta' f_B, \qquad M_A(t_0) = M_{A_0},$$

$$\dot{M}_B = -\beta M_B - \alpha M_A \alpha' f_A, \qquad M_B(t_0) = M_{B_0},$$

$$\dot{C}_A = \beta M_B (1 - \beta') v_B, \qquad C_A(t_0) = 0,$$

$$\dot{C}_B = \alpha M_A (1 - \alpha') v_A, \qquad C_B(t_0) = 0,$$

$$0 \le \alpha(t) \le \bar{\alpha}; \qquad 0 \le \alpha'(t) \le 1; \qquad \alpha(t),\ \alpha'(t) \text{ piecewise continuous.}$$

Here there are four control variables M_A, M_B, C_A and C_B, representing missiles in country A and B and casualties in country A and B, respectively. The four equations of motion summarize the evolution of a missile war between A and B, starting at time t_0 and ending at time t_1. At war outbreak each country has a certain number of missiles, given as M_{A_0} and M_{B_0}, respectively, in A and B, and no casualties. Country A fires its missiles at the rate $\alpha(t)$, so αM_A represents the number of A missiles launched at time t. Similarly βM_B represents the number of B missiles launched at time t. Missiles can be targeted at either enemy missiles (counterforce) or enemy cities (countervalue), and $\alpha'(t)$ and $\beta'(t)$ represent the counterforce proportions used by A and B, respectively, at time t. Thus $\beta M_B \beta'$ represents the number of B counterforce missiles launched at any time. The effectiveness of B missiles against A missiles is given by $f_B(t)$, the number of A missiles destroyed per B counterforce missile at time t. Similarly, $f_A(t)$ measures the number of

B missiles destroyed per A counterforce missile at time t. The countervalue effectiveness $v_B(t)$ measures the number of casualties inflicted in A per B countervalue missile at time t, while $v_A(t)$ measures the number of casualties inflicted in B per A countervalue missile at time t.

According to the first equation of motion, the number of A missiles decreases from its initial value of M_{A_0} because of two factors. One is the decisions of A to fire missiles, summarized by $-\alpha M_A$. The other is the decisions of B to fire counterforce missiles, which then destroy A missiles, summarized by $-\beta M_B \beta' f_B$.[†] Similarly, the two terms in the equation for $\dot{M}_B$ represent missiles launched by B and missiles destroyed by A counterforce missiles, respectively.

According to the third equation of motion, the number of casualties in A increases from its initial level of zero because of decisions by B to fire countervalue, summarized by $\beta M_B(1 - \beta')v_B$, where $1 - \beta'$ is the countervalue proportion, the proportion of B missiles targeted at A cities. The last equation then determines the rate of increase of B casualties as $\alpha M_A(1 - \alpha')v_A$.

The control variables $\alpha(t)$ and $\alpha'(t)$ summarize the rate of fire and targeting decisions of country A at time t. According to (24) the rate of fire can range between zero, which means holding missiles in reserve, and a given finite maximum rate $\bar{\alpha}$. The targeting decision is summarized by the counterforce proportion of α', which can range between zero, meaning firing pure countervalue, and unity, meaning firing pure counterforce. Over time both control variables can change, but they must be piecewise continuous functions of time.

The objective function in (24) is the payoff function for country A, which is to be maximized by choice of time paths

---

[†] Note that all these variables are time dependent. Explicit time dependencies have been omitted in (24) for clarity of exposition.

for the A rate of fire and counterforce proportion. This payoff function depends on the outcome of the war, summarized by the missiles and casualties in each of the countries at the time of war termination t_1.

Assuming the B strategy is given, the solution to the control problem in (24) entails switching strategies for both the rate of fire and the targeting strategies.† The solutions are

$$\alpha^*(t) = \begin{cases} \bar{\alpha} & \text{if} \quad t_0 \leq t \leq \tau \\ 0 & \quad\quad \tau < t \leq t_1 \end{cases}$$

$$\alpha'^*(t) = \begin{cases} 1 & \text{if} \quad t_0 \leq t \leq \tau' \\ 0 & \quad\quad \tau < t \leq t_1 \end{cases} \tag{25}$$

According to these solutions, the war always starts with counterforce targeting $(\alpha'^* = 1)$ at the maximum rate $(\alpha^* = \bar{\alpha})$, in order to reduce the weight of the enemy counterattack. It also always ends with countervalue targeting $(\alpha'^* = 0)$ at the zero rate $(\alpha^* = 0)$, holding missiles in reserve and threatening enemy cities in order to obtain a desired outcome. The middle stage is the critical one in terms of casualties inflicted in country B. If $\tau < \tau'$, so the rate strategy switches before the targeting strategy, then by the time A starts targeting B cities it has already stopped firing its missiles. In this case, then there are no casualties inflicted in B. If, however, $\tau' < \tau$, then in the time interval between the switching times

---

†The solutions are obtained using the maximum principle. Here there are four shadow prices, representing the value to A of retaining one of its own missiles, the value to A of destroying a B missile, the value of A of inflicting a B casualty, and the value to A of preventing one of its own casualties.

A is firing countervalue at the maximum rate, thereby inflicting casualties in B.

This type of analysis thus suggests both rapid shifts in optimal strategy and a particular form taken by these shifts in rate and target strategies. It further suggests that casualties are likely to be inflicted, if at all, only in the middle phase of the war. The basic model has also been used for analyses of war initiation, war termination, and arms control. In the application of arms control it is shown that in the missile plane (M_A, M_B) there exists a cone of mutual deterrence, for which each country deters the other with sufficiently large numbers of missiles on both sides. If the configuration of missiles implies a point well within the cone, then there are opportunities for unilateral or bilateral arms control measures [11].

VIII. CONCLUSION

The last several sections have presented some examples of microeconomic, macroeconomic, and other problems of economizing. While in these examples the microeconomic problems were static problems of programming and the macroeconomic problem was a dynamic one of control, there are examples of dynamic microeconomic problems and static macroeconomic problems. An example of the former is the problem of an optimal path of capital accumulation, including investments in research and development, by the firm. An example of the latter is the problem of choosing a point on the Phillips curve representing possible combinations of inflation and unemployment rates.

The static and dynamic problem of economizing have also been applied to other areas or issues, including, at the microeconomic level, the education and employment decisions of a worker, the inventory and employment decisions of a firm, the activities of a union, and the activities of a collective farm. At the macroeconomic level this approach has been applied to problems of economic stabilization and regulation, fiscal and monetary policy, international borrowing, and the allocation of investment among different sectors of the economy. The economizing approach can also be applied to areas not traditionally identified as part of economics but nevertheless involving the allocation of scarce resources, as indicated in Sections VI and VII on science policy and military strategy, respectively. Other areas that could be treated are educational planning and transportation policy.

The tools of economic analysis, specifically the methodology of the static and dynamic economizing problems, thus have been applied to a wide variety of areas. They promise to continue to provide useful analyses of both traditional and newer problems of allocating scarce resources among competing ends.

REFERENCES

1. M.D. INTRILIGATOR, "Mathematical Optimization and Economic Theory." Prentice-Hall, Englewood Cliffs, New Jersey, 1971.
2. P.A. SAMUELSON, "Foundations of Economic Analysis." Harvard University Press, Cambridge, Massachusetts, 1947.
3. F.P. RAMSEY, Economic J. 38, 543-559 (1928).
4. K. SHELL, "Essays on the Theory of Optimal Economic Growth." M.I.T. Press, Cambridge, Massachusetts, 1967.
5. K.J. ARROW, "Applications of Control Theory to Economic Growth", "Mathematics of the Decision Sciences", Part 2. Amer. Math. Soc., Providence, Rhode Island, 1968.
6. K. SHELL, "Mathematical Systems Theory and Economics", (H.W. Kuhn and G.P. Szegö, eds.). Springer-Verlag, Berlin and New York, 1969.
7. M.D. INTRILIGATOR, and B.L.R. SMITH, Amer. Econ. Rev. 56, 494-507 (1966).
8. R.H. BOLT, W.L. KOLTUN, and O.H. LEVINE, Science 148, 918-928 (1965).

9. M.D. INTRILIGATOR, "Security in a Missile War." Security Studies Project, University of California, Los Angeles, 1967.

10. M.D. INTRILIGATOR, Orbis 11, 1138-1159 (1968).

11. M.D. INTRILIGATOR, J. Pol. Econ. 83, 339-353 (1975).

MODERN AEROSPACE SYSTEMS

RANDALL V. GRESSANG

Flight Control Division
Air Force Flight Dynamics Laboratory
Wright-Patterson Air Force Base, Ohio

DEMETRIUS ZONARS

Air Force Flight Dynamics Laboratory
Wright-Patterson Air Force Base, Ohio

I. INTRODUCTION

Aerospace vehicle analysis and synthesis present fertile areas for the development and application of optimization techniques. Well-developed physical and mathematical models exist for many aspects of flight vehicle systems, and definite performance criteria can often be stated. This allows well-defined optimization problems to be posed for many aerospace vehicle systems. At the same time, the requirement for continually improving vehicle performance has led to requirements for improved design procedures and more accurate, but more complicated, system models. Optimization techniques hold the promise of providing design methods that will accommodate the more complicated models, and yield systems possessed of superior performance.

This chapter summarizes attempts to apply these optimization methods in several areas. The applications are divided into trajectory, performance, or tactics applications; tracking and attitude control applications; and vehicle and structural design problems. Considered specifically are steepest descent methods for trajectory synthesis, the application of differential games to air-to-air tactics, aircraft active control system design using quadratic optimal control, optimal human operator models in manual control systems, automated aircraft preliminary design, and optimal structural element sizing.

The objective of this chapter is to survey various optimization techniques that have been used or proposed for solving problems associated with atmospheric flight. For this survey, the problems will be classified into three types: trajectory, performance, or tactics problems; tracking and attitude control problems; and vehicle and structural design problems. This classification corresponds to considering the overall objectives of the vehicle, the control necessary to implement these overall objectives, and the details of a physical realization of the vehicle. Similarly, the optimization techniques employed vary from variational techniques for systems of nonlinear differential equations, to linear quadratic Gaussian optimal control theory, to mathematical programming and variational techniques for nondynamic systems.

The following section is concerned with applying the calculus of variations techniques to determining trajectories that optimize various performance criteria, and applications of differential game theory to air-to-air combat tactics. The third section considers mainly problems related to flight control and stability augmentation system design. Included in this section are applications of optimal control theory to modeling the man-machine system consisting of a pilot flying an aircraft under manual control. The succeeding section considers automating the aircraft preliminary design process to obtain an optimum configuration and the determination of optimum structures. The last section is a summary.

II. PERFORMANCE AND TACTICS OPTIMIZATION

A. Performance Optimization

Performance optimization is concerned with determining flight paths and trajectories for specified vehicles that maximize or minimize quantities such as range, payload, fuel consumption, or time to reach a specified altitude. A coordinate system commonly used in these problems is an earth-centered system, rotating with the earth, in which case the vehicle trajectory is determined by the vector differential equations

$$\dot{x} = v,$$

$$\dot{v} = \frac{F(x, v, u)}{m} - 2\omega_p \times v - \omega_p \times (\omega_p \times x). \quad (1)$$

Here x is the position vector specifying position of the vehicle, v the velocity of the vehicle, ω_p a vector specifying the earth's precession rate, m the vehicle's mass, F the vector giving the gravitational, aerodynamic, and control forces on the vehicle, and u a vector of control inputs. If the vehicle's mass can vary significantly, or fuel consumption is important, an additional differential equation for the mass change or fuel flow would be added. The control inputs would be some or all of the variables angle-of-attack, sideslip angle, bank angle, thrust magnitude, and two angles specifying the orientation of the thrust.

Even if the earth-centered coordinate system is not used, the vehicle dynamics are still described by six or seven non-linear differential equations. Constraints on the values of the state variables may be imposed throughout the entire trajectory, or only at the end of the trajectory. One of the trajectory constraints may be used to determine the end point of the trajectory. A constraint used to determine the final time of the trajectory is usually called a "cutoff condition". The quantity to be maximized or minimized is also usually expressed as a function of the state variables at the terminal time.

The optimization problem is then to maximize (or minimize) the scalar function

$$\phi = \phi(x(T), T) \quad (2)$$

subject to the constraints

$$\psi = \psi(x(T), T) = 0 \quad (3)$$

and

$$\dot{x}(t) = f(x(t),\ u(t)), \qquad t \in [t_0,\ T],$$
$$x(t_0) = x_0, \tag{4}$$

where $u(t)$ is a vector of piecewise constant (or square integrable) functions, defined on $[t_0,\ T]$. The final time T is determined by the equation

$$\Omega = \Omega(x(T),\ T) = 0. \tag{5}$$

This type of problem has been successfully solved using a variational steepest descent method, as outlined below.

The variational steepest descent method is described more fully by Hague [1], Mobley and Vorwald [2], and Hague and Glatt [3]. This method assumes that a nominal control history, with associated nominal trajectory, is known, and that the nominal trajectory satisfies the "cutoff condition". Then the control perturbations are determined that show the greatest improvement in the performance functional, while simultaneously removing a fixed amount of the error in the terminal constraints and keeping a norm on the control perturbations below a fixed upper bound. This process is then repeated until the desired accuracy is achieved.

The norm on the control perturbations is conveniently taken to be the inner product

$$\|\delta u\|^2 = \int_{t_{\sigma\sim 0}}^{T} \delta u'(t) W(t)\ \delta u(t)\ dt, \tag{6}$$

where W is an arbitrary symmetric positive definite matrix. Use of this inner product as the norm implies that $L_2[t_0,\ T]$, the space of functions square integrable on the interval $[t_0,\ T]$, is the space of admissible control perturbation. The effect of a control perturbation on the final state is determined by linearizing the differential equation about the nominal trajectory, and defining the adjoint system to the linearized equation to be

$$\dot{\lambda}(t) = -F'(t)\lambda(t), \tag{7}$$

where the linearized state equation is

$$\delta\dot{x}(t) = F(t)\,\delta x(t) + G(t)\,\delta u(t). \tag{8}$$

Here the prime denotes transpose, and δ denotes a perturbation.

From the identity

$$\frac{d}{dt}\{\lambda'\,\delta x\} = \lambda' G\,\delta u \tag{9}$$

and the three different sets of boundary conditions at $t = T$ given by

$$\lambda(T) = \left.\frac{\partial\phi}{\partial x}\right|_T = \lambda_\phi(T), \tag{10}$$

$$\lambda(T) = \left.\frac{\partial\Omega}{\partial x}\right|_T = \lambda_\Omega(T), \tag{11}$$

$$\lambda(T) = \left.\frac{\partial\psi}{\partial x}\right|_T = \lambda_\psi(T), \tag{12}$$

the changes in cost functional, "cutoff condition", and constraints due to control perturbations can be determined to be

$$\delta\phi = \int_{t_0}^{T} \lambda_\phi' G\,\delta u\,dt + \lambda_\phi'(t_0)\,\delta x(t_0) + \dot{\phi}(T)\,\Delta T, \tag{13}$$

$$d\Omega = \int_{t_0}^{T} \lambda_\Omega' G \, \delta u \, dt + \lambda_\Omega'(t_0) \, \delta x(t_0) + \dot{\Omega}(T) \, \Delta T, \tag{14}$$

$$d\psi = \int_{t_0}^{T} \lambda_\psi' G \, \delta u \, dt + \lambda_\psi'(t_0) \, \delta x(t_0) + \dot{\psi}(T) \, \Delta T. \tag{15}$$

In these equations, ΔT is the perturbation in the final time.

From the definition of the "cutoff condition",

$$d\Omega = 0, \tag{16}$$

therefore ΔT can be eliminated by using

$$\Delta T = -\frac{1}{\dot{\Omega}(T)} \left\{ \int_{t_0}^{T} \lambda_\Omega' G \, \delta u \, dt + \lambda_\Omega'(t_0) \, \delta x(t_0) \right\}. \tag{17}$$

The control perturbations that maximize the change in the cost functional, subject to specific changes in constraints and satisfying the cutoff condition, are determined by adding the constraints with Lagrange multipliers, and then performing an unconstrained maximization of the Lagrangian

$$\begin{aligned} L = & \int_{t_0}^{T} \lambda_{\phi\Omega}' G \, \delta u \, dt + \lambda_{\phi\Omega}'(t_0) \, \delta x(t_0) \\ & + \nu \left\{ \int_{t_0}^{T} \lambda_{\psi\Omega}' G \, \delta u \, dt + \lambda_{\psi\Omega}'(t_0) \, \delta x(t_0) \right\} \\ & + \mu \int_{t_0}^{T} \delta u'(t) \, W(t) \, \delta u(t) \, dt. \end{aligned} \tag{18}$$

Here ν and μ are the Lagrange multipliers, and

$$\lambda_{\phi\Omega} = \lambda_{\phi} - \frac{\dot{\phi}(T)}{\dot{\Omega}(T)} \lambda_{\Omega}, \tag{19}$$

$$\lambda_{\psi\Omega} = \lambda'_{\psi} - \frac{\dot{\psi}(T)}{\dot{\Omega}(T)} \lambda_{\Omega}. \tag{20}$$

Performing this minimization, the optimum control perturbation is then given by

$$\delta u = \pm W^{-1} G' [\lambda_{\phi\Omega} - \lambda'_{\phi\Omega} I^{-1}_{\psi\psi} I_{\psi\phi}]$$

$$\times \left(\frac{\|\delta u\|^2 - d\beta' I^{-1}_{\psi\psi}\, d\beta}{I_{\phi\phi} - I'_{\psi\phi} I^{-1}_{\psi\psi} I_{\psi\phi}} \right)^{\frac{1}{2}} + W^{-1} G' \lambda'_{\psi\Omega} I^{-1}_{\psi\psi}\, d\beta, \tag{21}$$

where

$$I_{\psi\psi} = \int_{t_0}^{T} \lambda'_{\psi\Omega} G W^{-1} G' \lambda_{\psi\Omega}\, dt, \tag{22}$$

$$I_{\psi\phi} = \int_{t_0}^{T} \lambda'_{\psi\Omega} G W^{-1} G' \lambda_{\phi\Omega}\, dt, \tag{23}$$

$$I_{\phi\phi} = \int_{t_0}^{T} \lambda'_{\phi\Omega} G W^{-1} G' \lambda_{\phi\Omega}\, dt, \tag{24}$$

and

$$d\beta = d\psi - \lambda'_{\psi\Omega}(t_0)\, \delta x(t_0). \tag{25}$$

The negative sign in the expression for δu corresponds to minimizing the performance functional, and the positive sign corresponds to maximizing.

By extending the definition of the inner product of the control perturbations to include discrete parameters, the effects of vehicle staging and configuration changes can be included in the variational steepest descent procedure. Hague and Glatt [3] outlines the modifications that must be made to incorporate staging and finite parameters.

An example of the application of the variational steepest descent method to aircraft performance optimization is given by the minimum time climb records of the F-4B aircraft. Fig. 1 shows the actual flight paths and calculated optimum flight paths for two F-4B time-to-climb record flights. It is readily seen that they correspond closely. Fig. 2 shows the flight path specified in the F-4B flight manual, the calculated minimum time flight path, and an attempt by Marine Col. Yunck to fly the optimal path. The optimal path and the attempt to fly the optimal path both resulted in a 23% decrease in the time required to reach the given altitude. Since no aircraft modifications were required to achieve this improved performance, the aircraft performance could be markedly increased just by flying it in a manner indicated by the optimization program. Further details of the F-4B performance optimization may be found in the work of Landgraf [4].

B. Air-To-Air Tactics

Since the first dogfights over the western front in World War I, what constitutes a good fighter aircraft and what are good fighter tactics have been questions whose answers are elusive. Most attempts to resolve these questions have used a trial-and-error approach, slightly modifying an existing aircraft or tactic, or designing the next aircraft to defeat the last adversary, and frequently the attempts have been unsuccessful. Similarly, the majority of fighter pilots have been unsuccessful in defeating other aircraft in air-to-air combat with most "kills" having been made by a small percentage of all pilots. Of some 45,000 pilots in World Wars I and II and the Korean conflict,

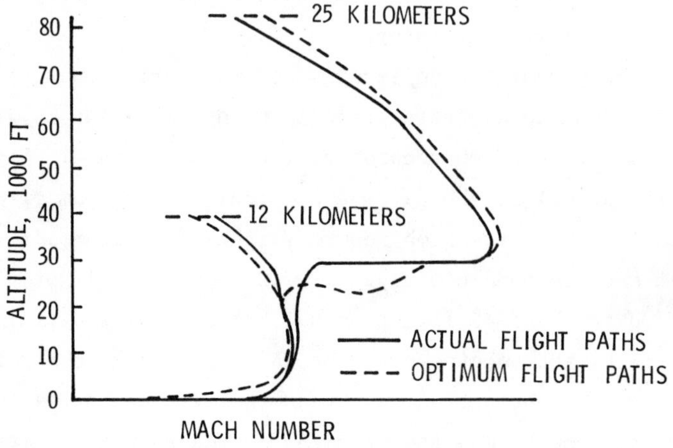

Fig. 1. Calculated Optimum Flight Path (dashed curves) and actual flight path (solid curves) for two F-4B time-to-climb record flights (From Hague and Glatt, [3]).

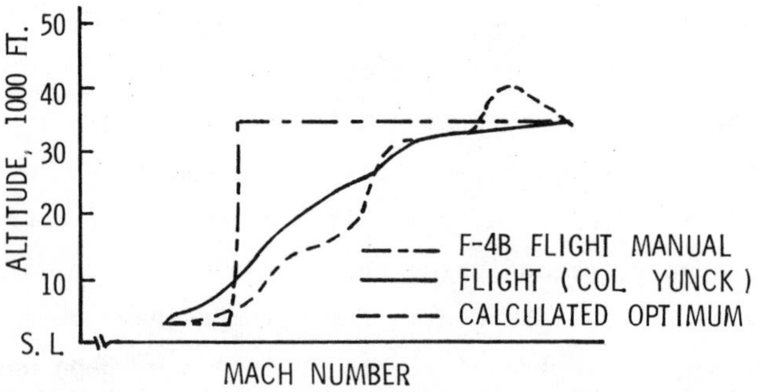

Fig. 2. Flight Path Comparison for F-4B Flight Manual (- • - •), flight by Col. Yunck (solid curve), and calculated optimum (dashed curve) (From Hague and Glatt [3]).

only about 1300 became Aces, with five or more aerial kills. This is about 3%. This section will survey the beginnings of an attempt to characterize what makes an aircraft a superior fighter, and what are superior tactics, by modeling the air-to-air engagement using the theory of differential games.

In a zero sum differential game, a dynamic system of the form

$$\dot{x} = f(x, u, v, t); \qquad x(t_0) = x_0 \tag{26}$$

is assumed, where x is an n-dimensional state vector, u the pursuer's (scalar) control, and v the evader's (scalar) control. In addition, a performance functional

$$J = \phi(x(t_f), t_f) + \int_{t_0}^{t_t} L(x, u, v, t)\, dt \tag{27}$$

and a terminating surface

$$\psi(x(t_f), t_f) = 0 \tag{28}$$

are assumed. The goal is to find controls u^* and v^* such that

$$J(u^*, v) \leq J(u^*, v^*) \leq J(u, v^*). \tag{29}$$

If such u^* and v^* exist, (u^*, v^*) is called a "saddle point" of the game, and $J(u^*, v^*)$ is the "value" of the game. The existence of the saddle point is dependent on the condition

$$\min_{u \in U} \max_{v \in V} J(u, v) = \max_{v \in V} \min_{u \in U} J(u, v) \tag{30}$$

holding, which is called the "minimax condition", where U and V are the sets of admissible control values for pursuer and evader.

Necessary conditions that u^* and v^* must satisfy can be derived using the Hamiltonian

$$H(t, x, \lambda, u, v) = \lambda' f + L. \tag{31}$$

The conditions are that

$$H^* = \max_v \min_u H = \min_u \max_v H, \tag{32}$$

where λ is an n-dimensional adjoint vector satisfying

$$\dot{\lambda}' = -H_x \tag{33}$$

with the transversality conditions

$$\lambda'(t_f) = \phi_x(t_f). \tag{34}$$

If f, L, and H each separate into a part depending only on u, and a part depending only upon v, then the minimax condition on H can be satisfied. However, satisfying the minimax condition for H does not imply that it is satisfied for J, so candidate saddle point solutions derived from the necessary conditions must be verified using sufficiency conditions. In addition, frequently u and v appear linearly in H, so that H cannot be maximized or minimized with respect to u or v, and singular arcs must be considered.

The necessary conditions outlined above result in a two-point boundary value problem that must be solved for u and v. If this boundary value problem can be solved only for $u(x_0, t)$ and $v(x_0, t)$, the solution is termed "open loop". If, however, the solution is given in the form $u = u(x, t)$ and $v = v(x, t)$, it is termed "closed loop". In a differential game, only closed-loop solutions are of significant interest, as simple examples exist which show that if one player knows that the other player is using an open-loop optimal strategy, by playing suboptimally (i.e., not his own open-loop strategy), the first player can improve his outcome.

The preceding discussion has basically been of the type of differential game called a "game of degree", in that an implicit assumption was made that the terminating surface could be reached. In a game of kind (Isaacs, [5]) the object of the pursuer is to reach the terminating surface, and the object of the evader is to prevent the pursuer from reaching the terminating surface. The payoff for a game of kind thus depends only on whether the terminal surface can be reached in the time alloted, and the situation is similar to controllability in a control system.

Let ν be an outward normal to the terminating surface

$$\psi(x(t_f),\ t_f) = 0. \tag{35}$$

Then for capture to occur,

$$\nu' f(x,\ u,\ v,\ t) < 0 \tag{36}$$

Thus the sign of the quantity

$$\min_u \max_v \nu' f(x,\ u,\ v,\ t) \tag{37}$$

can be used to subdivide the terminal surface into a part where capture can never occur, a part where capture can always be forced, and a neutral boundary between these areas, called the "boundary of the useable part". The "barrier" is then defined to be the surface of state positions which lead to a neutral outcome, i.e., for optimal play, the trajectory starting at a state on the barrier will terminate on the boundary of the useable part. The barrier thus indicates which regions of the state space are favorable to the pursuer, and which are favorable to the evader. It should be noted that a barrier need not be a closed surface that divides the state space into two regions.

Othling [6] has formulated air-to-air combat problems as zero sum games of degree. His most general problem considered one aircraft attempting to close with another in the vertical plane, with the performance functional being the range between the aircraft at a fixed terminal time t_f. Each aircraft was modeled as a point mass, the equations being

$$\dot{x} = V \cos \gamma, \tag{38}$$

$$\dot{h} = V \sin \gamma, \tag{39}$$

$$\dot{V}_x = (G/W)[(T-D) \cos \gamma - L \sin \gamma], \tag{40}$$

$$\dot{V}_h = (G/W)[(T-D) \sin \gamma - L \cos \gamma] - G, \tag{41}$$

where

$$\gamma = \mathrm{ARCTAN}(V_h/V_x), \tag{42}$$

$$D = \frac{1}{2} \rho V^2 S C_D, \tag{43}$$

$$L = \frac{1}{2} \rho V^2 S C_L, \tag{44}$$

and the control is taken to be the lift coefficient C_L. Here V is the aircraft velocity, γ the flight path angle, T the thrust, D the drag, L the lift, G the acceleration due to gravity, W the aircraft weight, V_h the vertical velocity, V_x the longitudinal velocity, ρ the atmospheric density, S the wing area, C_D the drag coefficient, and C_L the lift coefficient. Certain other assumptions, such as constant thrust, are made which limit the problem to small altitude and velocity changes. He was unable to determine a closed-loop solution to the most general game he posed. However, by finding closed-loop solutions to a series of simplified cases, he was able to determine a suboptimal closed-loop control for pursuer and evader that compared well with open loop solutions.

Considering air-to-air combat problems as zero sum games of kind, both Lynch [7] and Miller [8] have considered problems related to whether an attacker can reach a firing position, or the evader successfully evades. Beginning with simple two-dimensional games such as Isaacs' homicidal chauffeur game and game of two cars [5], they consider the effect on the barrier of progressively improving the realism of the model used, and evaluate the effect on the barrier of aircraft parameters such as acceleration along the velocity vector, weapon system range, turning acceleration, and thrust-to-weight ratio. Lynch found that increasing the thrust-to-weight ratio resulted in the greatest performance improvements in his problems, and Miller found that increases in acceleration along the velocity vector and maximum lift coefficient, or a decrease in wing loading, were particularly effective in improving performance. In his study, Miller included various angle-off conditions to represent machine gun type armament.

In addition to the application of zero sum differential game theory discussed above, Leatham [9] investigated the application of nonzero sum differential game theory to combat problems. In a nonzero sum differential game, the performance functionals of the two players are not exactly opposite and conceptual difficulties appear that are not present in zero sum differential games. Leatham used a Hamilton-Jacobi partial differential equation approach, and considered interceptor-penetrator problems, with two or three players.

To date, the differential game approach to air-to-air combat problems has produced mainly verification that known tactics and figures of merit are "nearly optimal". Because of the nonlinearity and complexity of the problems, and the difficulty in determining closed-loop solutions, considerable work remains before differential game theory provides an accepted basis for evaluating fighter aircraft and tactics. The difficulty in determining closed-loop solutions could be circumvented by developing algorithms for determining open-loop solutions efficiently in real time.

III. TRACKING AND ATTITUDE CONTROL

Since the Wright 1903 Flyer, aircraft have been designed not to be inherently dynamically stable, but to be able to be stabilized (controlled) by a pilot or autopilot. The main part of this section will be concerned with applications of modern control theory to designing systems to stabilize an aircraft (allow it to maintain a precise attitude and to follow commands) and to analyze the interaction of the pilot and aircraft. In a manned aircraft, the pilot is an important, complex part of the system, who can be described very imprecisely in engineering terms. His abilities and limitations strongly influence the rest of the vehicle.

A. Aircraft Active Control Systems

Demands for high performance, long life, and low cost in aircraft, coupled with the capability of routinely performing lengthy calculations using a digital computer, have resulted in more attention being paid to benefits that could be derived from complex control systems. Vehicles in which flight control receives equal emphasis with aerodynamics, structures, and propulsion in preliminary design are termed "control configured vehicles" (CCV). The requirements that the control system may be expected to satisfy, in addition to the classical requirements of stabilization, are load alleviation and mode stabilization, ride control, flutter mode control, augmented stability, maneuver load control, and gust alleviation. These additional requirements have the purpose of reducing fatigue on the aircraft, improving crew comfort, reducing aeroelastic effects so airspeed can be increased, reducing drag and control surface size while maintaining adequate stability margins, minimizing wing-bending stresses during maneuvers, and reducing the effects of atmospheric turbulence.

In addition to the control configured vehicle requirements, the control system also must satisfy handling qualities requirements, which arise because the control system must be considered as part of a man/machine system. Handling qualities represent a basically empirical data base in terms of what aircraft dynamic responses have been found desirable by pilots. For control system design using optimization techniques, the handling quality data can be summarized either in the form of a linear constant coefficient differential equation model, whose dynamics represent good handling qualities, or in the form of a pilot rating function. The pilot rating function expresses pilot opinion (usually using the Cooper-Harper scale) as a function of system performance. To date, the handling qualities model [10, 11] (differential equation) approach has been the main technique used, with few applications of the pilot rating function approach [12].

To design a control system that will satisfy the above requirements, the approach has been to develop a linear perturbation model of the aircraft, and to represent the requirements by a quadratic cost functional. The techniques of linear quadratic Gaussian optimal control theory [13] are then used to design multiple sensor, multiple controller feedback control laws. The main application of optimal control theory is to provide a systematic rather than a trial-and-error method for determining the feedback controls. The actual quadratic cost functional used is not usually important; frequently several are used, and the cost functional that results in a control best satisfying the requirements listed above is selected.

The aircraft is modeled by a linear perturbation stochastic differential equation [14-16], for each flight condition of interest, of the form

$$dx = Fx\,dt + Gu\,dt + B\,d\eta, \tag{45}$$

where x is the state vector, whose elements might include aircraft structural modes, sensor dynamics, a handling qualities model, and a dynamic model for gusts and disturbances [17] as well as the aircraft rigid body states, u is a vector of control inputs, and η a vector of uncorrelated, unity variance Brownian motion processes. Matrices F, G, and B specify the system dynamics, control distribution, and disturbance distribution, respectively. The sensor dynamics are assumed to be included in the state equation, and the sensor outputs (measurements) are given by

$$y = Hx, \tag{46}$$

where y is the vector of measurements, and H a matrix relating measurements to states.

The quadratic cost functional is formulated as

$$J = E[x'Qx + x'Su + u'S'x + u'Ru] \tag{47}$$

where $E[\cdot]$ is an operator denoting the expectation of a random variable. Equivalently, a vector r, composed of responses that are to be minimized, is defined by

$$r = Mx + Du, \tag{48}$$

and the quadratic cost functional is then defined as

$$J = E[r'Qr] = E[\mathrm{tr}[Qrr']]. \tag{49}$$

The control law is specified to be of the form

$$u = Ky. \tag{50}$$

This form of control law is adopted as it results in a physically realizeable control system, which can be constructed in a simple, reliable manner. The optimization problem is defined to be the minimization of J as a function of the elements of K, subject to the condition that the stochastic differential equation with the control be asymptotically stable.

Two approaches have been used on this problem, either to use a gradient numerical technique or to derive a necessary condition that K must satisfy, and then solve the necessary condition for K. The gradient search techniques are based on the fact that if K is known, the cost functional J and its gradient with respect to elements of K can be calculated effectively.

Given K, the value of the cost functional J can be calculated by solving the linear matrix equation

$$[F + GKH]P + P[F + GKH]' + BB' = 0 \tag{51}$$

for $P = E[xx']$. This equation has a solution if the eigenvalues of the matrix $F + GKH$ lie in the left half of the complex plan (which is the stability constraint), and the solution can be calculated efficiently using an infinite series expansion [18]. The cost functional J is then calculated from

$$J = \mathrm{tr}[Q(M + DKH)P(M + DKH)']. \tag{52}$$

In a similar way, the gradient of J with respect to K can be computed by solving the linear matrix equations

$$[F + GKH]\frac{dP}{dk_{ij}} + \frac{dP}{dk_{ij}}[F + GKH]' + G\frac{dK}{dk_{ij}}HP + P[G\frac{dK}{dk_{ij}}H]' = 0, \tag{53}$$

where k_{ij} is an element of K, and then using

$$\frac{dJ}{dk_{ij}} = \mathrm{tr}[Q(M + D\frac{dK}{dk_{ij}}H)P(M + DKH)' + Q(M + DKH)P(M + D\frac{dK}{dk_{ij}}H)' + Q(M + DKH)\frac{dP}{dk_{ij}}(M + DKH)']. \qquad (54)$$

Even though J and its gradient can be computed efficiently, the problems remain of where to start whatever gradient technique is used, and what Q to use in the cost functional. These questions are answered by solving a series of optimal state feedback control problems for several differential Q matrices, until a Q is found that results in the requirements being satisfied by the optimal state feedback control. These optimal control problems are set up and solved using the now classical matrix Riccati equation [Kwakernaak and Sivan, 19]. The optimal state feedback gain matrix K^* is then edited to yield a K for the initial iteration of the gradient algorithm. Both gradient and conjugate gradient techniques have then been used to minimize J as a functional of the elements of K.

An example of the application of the gradient technique is Poyneer's [20] design of lateral and longitudinal control systems for a CCV B-52. The requirements on the control system included ride control, maneuver load control, and load alleviation. Aircraft rigid body dynamics, structural mode damping, and handling qualities were also affected by the control system. The lateral system had five control surfaces (rudder, flaperon, inboard aileron, outboard aileron, and vertical canard), 10 measurements (those available to a conventional system, plus additional accelerometers and rate gyros), and was modeled by 37 states. Of 31 goals for the lateral system, 27 were satisfied. The longitudinal system also had five control surface sets (elevator, inboard aileron, outboard aileron, flaperon, and horizontal canards) and seven measurements. The longitudinal system met 36 out of 41 of the goals set for it. Both the lateral and longitudinal control systems were adequate for the intended use.

Heath [21] has extended the gradient approach for designing nondynamic feedback controls by deriving an expression for the gradient in the case where stochastic parameters appear in the stochastic differential equation modeling the aircraft. The object of including stochastic parameters in the aircraft model is to permit optimization of controls for more than one design point and to account for uncertainty in knowledge of the aircraft stability derivatives, etc. He then applied his gradient expression in designing a lateral stability augmentation system for an F-4 aircraft at two flight conditions, using an 11 state model, and assuming two controls (rudder and aileron) and four measurements (lateral acceleration, roll rate, yaw rate, and bank angle). The resulting system met the requirement of improving the handling qualities to be within specification.

For the necessary condition approach, consider the control problem formulated with the cost functional

$$J = E[x'Qx + u'Ru] \tag{55}$$

and plant stochastic differential equation

$$dx = Fx\,dt + Gu\,dt + B\,d\eta, \tag{56}$$

$$y = Hx. \tag{57}$$

The cost functional can be rewritten as

$$J = \mathrm{tr}[[Q + H'K'RKH]P], \tag{58}$$

where

$$[F + GKH]P + P[F + GKH]' + BB' = 0 \tag{59}$$

and

$$u = Ky. \tag{60}$$

Using these equations, the Hamiltonian

$$H = \operatorname{tr}[[Q + H'K'RKH]P] + \operatorname{tr}[S'[(F + GKH)P + P(F + GKH)' + BB']] \tag{61}$$

is formed, and the necessary conditions that K must satisfy are derived. These necessary conditions are

$$\frac{\partial H}{\partial S} = (F + GKH)P + P(F + GKH)' + BB' = 0, \tag{62}$$

$$\frac{\partial H}{\partial P} = S(F + GKH) + (F + GKH)'S + Q + H'K'RKH = 0, \tag{63}$$

$$\frac{\partial H}{\partial K} = [RKH + G'S]PH' = 0. \tag{64}$$

These conditions were first derived by Axsater [22] (time-varying plants) and Levine and Athans [23] (deterministic problems).

The feedback gain matrix K is then determined by solving the necessary conditions. The necessary conditions are solved numerically using a gradient algorithm, or some special purpose algorithms that have been proposed, such as Axsater's [22]. All of the algorithms require an appropriate initial choice for K.

The necessary condition approach has been used by Stein and Henke [10] to design a lateral control system for an F-4C aircraft, using a model having 20 states. The main requirement of the control system was to improve the aircraft handling qualities. The procedure was successful in designing a control system that met the requirements, with the satisfaction of the requirements being demonstrated by a handling qualities simulation experiment.

The feedback controls considered above have been fixed in form, in that the controls were required to be linear combinations of the measurements. If a dynamic compensator is allowed, and the controls are not required to consist solely of linear combinations of the measurements, the separation theorem of stochastic control [Wonham, 24] and Luenberger [25] observers can be used. Use of the separation theorem to design the controller, as the cascade of an optimal filter followed by an optimal state feedback control, would guarantee the best performance attainable by a linear feedback control for that cost functional. These types of control systems have not found wide acceptance, however, due to their complexity. To implement an optimal filter, as required by the separation theorem, would require as many states in the control system as were in the aircraft model. One example of using an observer in a control system (for taxiing an air cushion landing system aircraft) can be found in the work of Gressang [26]. In this case, the aircraft model was considerably simplified before the control system was designed.

In designing flight control systems, consideration must be given to the fact that the aircraft's dynamic characteristics can vary considerably over its flight envelope. The applications of linear quadratic optimal control to flight control design given above have all basically been concerned with only one flight condition, or a small number of flight conditions. In order to develop a practical control system, the "frozen point" flight control designs for specific flight conditions must be integrated in an appropriate fashion. One method of doing this is to try to determine one flight control system that gives acceptable performance over the whole range of anticipated flight conditions. Heath's [21] F-4 design is an example of this approach. In general, however, one feedback gain matrix does not give satisfactory performance over the range of anticipated flight conditions. Several gain matrices are required, for several regions of anticipated flight conditions, and the appropriate gains are selected by gain scheduling as a function of air data (such as dynamic pressure) or by use of an adaptive controller method.

B. Pilot-Vehicle Analysis

A manned aircraft is by definition a man-machine system, and cannot be completely analyzed or synthesized without considering the man. In the control system design procedures considered in the previous section, man was considered in an indirect way, through handling qualities models. In this section, the emphasis will be on developing mathematical models of human behavior in well specified tasks.

Since the 1940s, there have been developed effective analysis techniques and a supporting data base for modeling human operator response in manual control systems [27, 28]. The main analytical tools have been drawn from the frequency domain concepts of classical control theory. However, two of the concepts have been based on state space control theory and optimization, and will be described here.

First to be discussed will be an optimal pilot model developed at Bolt Beranek and Newman, Inc. [29]. The basic assumption of the model is that a well-trained, well-motivated pilot doing a well-defined task behaves optimally subject to his inherent human limitations. These limitations are considered to be:

(1) a time delay, representing cognitive, visual central processing, and neuromotor delays,

(2) "remnant" signals, divided into an observation noise to represent visual thresholds and divided attention, and a motor

noise to represent errors in executing intended movements,

(3) a "neuromuscular lag" to represent neuromuscular dynamics.

The model is based on a suboptimal solution to a control problem involving a time delay and observation noise [Kleinman, 30]. The pilot is represented by an optimal filter (Kalman filter), followed by an optimal predictor, and then an optimal state feedback control law. The control law is derived from a cost functional which weights control rates instead of control magnitude, and this introduces the neuromuscular lag. Fig. 3 is a block diagram showing the structure of the model.

In applying the model, the vehicle and actuator dynamics are first represented by the linear stochastic differential equation

$$dx(t) = Ax(t)\,dt + Bu(t)\,dt + G\,dw(t), \tag{65}$$

where $x(t)$ represents the n-dimensional vehicle state, $u(t)$ the pilot's (scalar) input to the vehicle, and $w(t)$ a Brownian motion process representing disturbances such as atmospheric turbulence. The system outputs are represented by the measurement equation

$$y(t) = Cx(t) + Du(t). \tag{66}$$

It is usually assumed that the pilot has available to him both the values of the displayed quantities and their first derivatives, but that he cannot determine higher derivatives or the integrals of displayed quantities. Also, the pilot can make observations only after a time delay τ representing the cognitive, visual central processing, and neuromotor delays. Therefore, the observations available to the pilot are given by

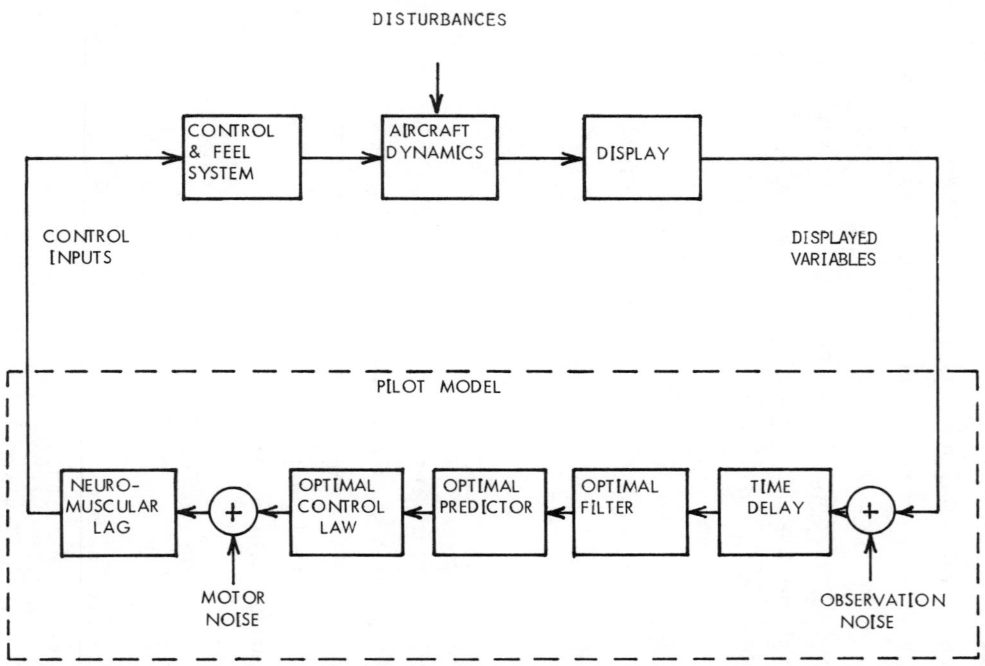

Fig. 3. Structure of Optimal Pilot Model.

$$y_p(t) = y(t - \tau) + v_y(t - \tau), \quad (67)$$

where τ typically has a value in the range 0.15 - 0.3 sec. and v_y a white-noise process representing observation noise and perception thresholds. Its standard deviation is discussed subsequently.

After the vehicle dynamics and the observations are modeled, a quadratic cost function of the form

$$J = E[y'Qy + R\dot{u}^2] \quad (68)$$

is assumed. The optimal control law is determined by solving the Riccati equation

$$A_0'K_0 + K_0A_0 + C_0'QC_0 - K_0B_0R^{-1}B_0'K_0 = 0, \quad (69)$$

where

$$B_0 = [0 \cdots 0 \; 1]', \quad (70)$$

$$A_0 = \begin{bmatrix} A & B \\ 0 & 0 \end{bmatrix}, \quad (71)$$

and

$$C_0 = [\, C \quad D \,]. \quad (72)$$

Then

$$\lambda = B_0'R^{-1}K_0, \quad (73)$$

and R is adjusted until the "neuromuscular lag" τ_N defined by

$$\tau_N = 1/\lambda_{n+1} \quad (74)$$

lies in the range 0.07 - 0.3 sec. The optimal gain ℓ is determined by

$$\ell_i = \tau_N \lambda_i , \qquad i = 1, \ldots, n \tag{75}$$

and the optimal control law, neuromuscular lag, and motor noise are represented by

$$du(t) = -\frac{1}{\tau_N} u(t)\, dt - \frac{1}{\tau_N} \ell\, \hat{x}(t)\, dt + dv_m(t), \tag{76}$$

where $\hat{x}(t)$ is the output of the optimal predictor, and $v_m(t)$ a Brownian motion process representing the motor noise.

The optimal predictor output $\hat{\chi}(t)$ is determined by the Kalman filter

$$\begin{aligned} \dot{\hat{\chi}}(t - \tau) = A_1 \hat{\chi}(t - \tau) + \textstyle\sum C_0' V_y^{-1} \\ \times [y_p(t) - C_0 \hat{\chi}(t - \tau)] \\ - B_0 \tau_N^{-1} \ell \hat{x}(t) \end{aligned} \tag{77}$$

and the optimal predictor by

$$\hat{\chi}(t) = \xi(t) + e^{A_1 \tau} [\hat{\chi}(t - \tau) - \xi(t - \tau)], \tag{78}$$

$$\dot{\xi}(t) = A_1 \xi(t) - B_0 \tau_N^{-1} \ell \hat{x}(t), \tag{79}$$

where $\sum$ is determined as the positive definite solution of the Riccati equation

$$A_1 \textstyle\sum + \sum A_1' + W - \sum C_0' V_y^{-1} C_0 \sum = 0, \tag{80}$$

where

$$A_1 = \begin{bmatrix} A & B \\ 0 & -\tau_N^{-1} \end{bmatrix}, \tag{81}$$

$$W = \begin{bmatrix} GG' & 0 \\ 0 & V_m \tau_N^{-2} \end{bmatrix}, \tag{82}$$

and

$$\hat{\chi}(t) = [\hat{x}(t), \hat{u}(t)]'. \tag{83}$$

The covariance of $\chi(t)$ is then given by

$$X = E[\chi(t)\chi'(t)]$$

$$= e^{A_1\tau} \sum e^{A_1'\tau} + \int_0^{\tau} e^{A_1\sigma} W e^{A_1'\sigma}\, d\sigma$$

$$+ \int_0^{\infty} e^{\bar{A}\sigma} e^{A_1\tau} \sum C_0' V_y^{-1} C_0 \sum e^{A_1'\tau} e^{\bar{A}'\sigma}\, d\sigma \tag{84}$$

with

$$\bar{A} = A_0 - B_0\lambda. \tag{85}$$

Then

$$E[y_i^2(t)] = (C_0 X C_0')_{ii} , \tag{86}$$

$$E[u^2(t)] = X_{n+1,n+1} , \tag{87}$$

and the model is complete when the covariances V_m and V_y of the motor noise and observation noise are specified.

The time delay has values typically in the range 0.15 - 0.3 sec. Using a value in this range for the time delay, the covariances V_{m} and V_y are then determined by requiring that

$$V_{\mathrm{m}} = \pi P_{\mathrm{m}} E[(\ell \hat{x}(t))^2] \tag{88}$$

and

$$(V_y)_{ii} = \pi P_{y_i} T(y_i^2) E[y_i^2], \quad i = 1, \ldots, n, \tag{89}$$

where P_{m} is a control-to-motor-noise ratio, typically having a value of 0.003, and P_{y_i} is a signal-to-noise ratio, typically having a value of 0.01 for single-axis tracking tasks, and $\pi = 3.14159 \cdots$. The term $T(y_i^2)$ is a describing function gain representing threshold, and if more than one axis of tracking is being performed, or the pilot's attention is being distracted, P_{y_i} may be raised to 0.02 or higher.

This pilot model appears to have great potential, as it includes within it a performance analysis of the man-machine system, and can be extended to vehicles having time-varying dynamics [Kleinman and Perkins, 31]. An example of the application of this pilot model can be found in Harvey's [32] study of air-to-air gunnery tracking. He used linearized equations for pitch tracking an evading aircraft. The evading aircraft was represented by a Gauss-Markov random process, and the attacking aircraft was represented by its short period dynamics and the dynamics of its lead computing optical sight. Geometric effects were included in the dynamics; however, the target range was assumed to be constant. Harvey made model parameter adjustments to evaluate tracking performance for three aircraft (the F-4, F-5, and A-7) at two ranges. These performance measures were matched with the results of a fixed base simulation of the same situation, and the agreement was found to be excellent using the final set of model parameters. Harvey found that the model results were sensitive to incorporating the correct thresholds on observations, and to having the correct weighting on rates in the cost functional.

The second application of optimization to pilot vehicle analysis, called "Paper Pilot", was conceived by Anderson [33] as a method of specifying handling qualities of unconventional aircraft. The basic assumption underlying Paper Pilot is that the pilot opinion rating of a particular aircraft configuration is a function of closed-loop aircraft performance and pilot workload, and the pilot adapts himself so as to obtain a minimum rating. (On the Cooper-Harper and other pilot opinion rating scales, a smaller rating implies a better aircraft.)

In performing a paper pilot analysis, only a specific class of aircraft performing a well defined task is considered. The pilot is represented by a fixed form pilot model appropriate to the closed-loop task [27, 28] and then a stochastic differential equation representing the pilot vehicle system is derived. This equation is of the form

$$dx = A(k)x\ dt + B(k)\ d\eta, \tag{90}$$

where x is the state of the pilot vehicle system, k a vector of pilot parameters, and $d\eta$ a Brownian motion process representing gust disturbances, etc. The time delay in the pilot model is represented by a Padé approximation.

The closed-loop system performance is then determined by solving the Liapunov matrix equation

$$A(k)P + PA'(k) + B(k)B'(k) = 0 \tag{91}$$

with

$$P = E[xx']. \tag{92}$$

Assuming that a function is known which gives pilot opinion as a function of performance and pilot parameters,

$$\text{pilot rating} = f(P, k). \tag{93}$$

The pilot parameters can now be determined by selecting k to minimize the pilot rating, subject to constraints on the elements of k given by human response theory. (Minimization is required as a smaller rating is a better rating.) In the Paper Pilot analyses performed to date, this minimization has been accomplished using a conjugate gradient algorithm. The gradient has been computed with the aid of the equation

$$A(k) \frac{dP}{dk_i} + \frac{dP}{dk_i} A'(k) + \frac{dA(k)}{dk_i} P + P \frac{dA'(k)}{dk_i}$$

$$+ \frac{d}{dk_i} [B(k)B'(k)] = 0. \tag{94}$$

Paper Pilot can be applied to flight control system design by including parameters other than the pilot parameters in the minimization. It can also be used to determine pilot parameters, closed-loop performance, and pilot ratings for proposed aircraft, before they are constructed or simulated. However, these applications require that pilot rating expressions be available for the tasks of interest. Most research on Paper Pilot to date has been concerned with establishing the validity of the concept, and determining appropriate pilot rating expressions. An example is the work of Arnold [34], who obtained data from a fixed-base simulation of the short period equations of motion of 14 different jet fighter type configurations. Using these data, he was able to develop a pilot rating expression and determine the form of pilot model that closely matched that data. Use of this pilot rating expression and pilot model form in a Paper

Pilot analysis then accurately predicted pilot rating, pilot parameters, and closed-loop pilot vehicle performance.

Pilot models can be used not only to predict closed-loop man-machine performance, and adjust flight control system parameter values, but also to decide between different control system or display structures. In this application, a pilot model is determined for each of the configurations being considered. This pilot model will reflect in its structure the changes in the task occasioned by the configuration differences. Then using the pilot model, a performance analysis can be done for the closed-loop man-machine system, and the differences in performance between the various configurations determined. At the same time, the assumptions underlying the model can be evaluated, so as to avoid unrealistic conclusions.

An example of this process is the work of Leondes and Rankine [35]. They considered the effect of pilot performance and tracking capability on the strafing accuracy of a tactical fighter. A linearized model was developed to relate aiming and flight path errors to weapon impact errors, and the aiming and flight path errors were determined from a closed loop model of the aircraft dynamics, gust disturbances, and pilot. Once the model had been developed, it was then used to determine analytically the improvement in strafing accuracy to be expected if conventional instruments were replaced by a heads up display (HUD).

C. Miscellaneous: Microwave Landing System Analysis, Active Landing Gear, Reliability

In this section two applications of linear optimal control theory and the possibility of using dynamic programming to design reliable flight controls will be briefly considered.

The first application concerns analyzing proposed microwave landing systems to determine required data rates and measurements

[35]. A "window" is defined in terms of the aircraft states, and approach performance is defined in terms of the probability of missing the window. The landing approach task is modeled by a stochastic differential equation, which incorporates aircraft dynamics, gust disturbances, and the guidance system. The flight control system is modeled as a state estimator followed by a state feedback control, which is optimized to minimize the probability of missing the window subject to constraints on the root mean square values of the control variables.

The system performance is then determined as a function of data rate by determining performance for continuous measurements and variations in gust environment, guidance system, etc., then repeating this performance determination with measurements at a specific data rate for the same variations. This determines a system degradation due to data rate, and allows an acceptable data rate to be chosen based on the minimum performance required of the system.

The above procedure has been applied to evaluating a scanning beam microwave landing guidance system, using DC-8 and CH-53A aircraft dynamics [Dillow *et al.*, 36]. The analytical approach described above was extremely successful, obtaining in a relatively short time information that had taken months to acquire for a single case using a Monte Carlo simulation. The approach has also been extended from glide slope tracking through flare to touchdown, by incorporating an equation for determining the mean value of the aircraft state. For this extension, only the DC-8 aircraft dynamics were used [Huber, 37].

The second control theory application is the design of an active landing gear control for vibration isolation during taxiing. The requirement is to reduce wing fatigue that arises from flexing induced by taxiing over rough runways and taxiways. The problem is formulated as an optimal feedback control problem, where the feedback controls are constrained to be linear combinations of the available measurements. A linear stochastic differential equation is derived to represent the aircraft and landing gear dynamics, and the disturbances introduced by runway unevenness and roughness. The measurements are considered to be noiseless, and the cost functional is selected to minimize wing flexure. Corsetti and Dillow [38] formulated this problem for several different landing gear models, and solved the resulting optimization problem using a Newton Raphson iterative procedure. They found that the stress on the wing could be made arbitrarily small by feeding back the appropriate displacements and accelerations of aircraft and landing gear masses.

The final comments of this section are on designing reliable systems through proper allocation of redundancy. This problem has basically not been addressed, except to calculate failure rates and mean time between failures for proposed systems. Two ideas, which look promising, but have not been extensively developed for application to flight controls, are to apply dynamic programming techniques to determine where redundancy is most valuable [Connors, 39], and to use optimal filtering techniques to determine if a failure has occurred and the flight control system should be reorganized to compensate for the failure [Maybeck, 40].

V. VEHICLE AND STRUCTURAL DESIGN

The increased complexity of modern aerospace vehicles, and the more precise analysis and synthesis methods employed to design them, have resulted in considerably more time and resources having to be devoted to vehicle design. Especially noteworthy is the much larger data base required for design, and the larger number of calculations that must be performed. Because of the increased amount of data that must be used, and the larger number of calculations, automation of design or computer-aided design becomes important. Not only can computer-aided design shorten the time and reduce the resources required for vehicle design, but it can permit more iterations of the design process to be performed, thus allowing more optimization of the design to be considered.

A. Aircraft Preliminary Design

In this section, automation of aircraft preliminary design will be treated by examining a specific computer program ODIN/MFV, which has that object [3]. The underlying concept of the program is to produce a computer-operated replica of the conventional preliminary design procedure, in a modular fashion so that improvements in various areas can be readily incorporated. The design is optimized by first determining a nominal design, and then searching about the nominal to determine an optimum.

The program is configured to include operational constraints and performance criteria from among landing and take-off performance, payload capability, maximum acceleration and lift coefficient maneuver limits, thermodynamic constraints, and economic constraints. It is organized as an executive program, which can call on various technology programs, a design optimization program, programs for reporting results graphically and numerically, and a data base for passing the results of one program to another.

The technology programs are for geometry, aerodynamics, propulsion, mass and volume, performance, aeroelasticity, and cost. The geometry program specifies a system of quadrilateral surfaces covering the vehicle's surface, and provides three-view and perspective drawings of the vehicle. The aerodynamic program is composed of separate programs for subsonic, supersoinic, and hypersonic flight. It uses the output of the geometry program, and computes, as required, hypersonic viscous and pressure forces, supersonic zero-lift wave drag, wave drage at lift, wetted areas, skin friction drag, and trend analyses. The propulsion program can calculate steady-state design and off-design performance for turbofan and turbojet engines. In the mass and volume program, statistics of past designs are used to estimate the weight and volume apportioned to the various subsystems. The weight and volume apportioned are based on regression functions of physical parameters. Three options are present in the performance program; either a simplified take-off and landing analysis, a segmented mission analysis, or a three-degree-of-freedom flight path optimization.

Because of computer time and core storage restrictions, the segmented mission analysis is most useful. The aeroelastic program calculates the spanwise aerodynamic forces and moments, and then assuming conventional wing structures, adjusts the wing section stiffness values until the resulting wing will not exceed the stress distributions allowed for the given load conditions. The final technology program, the economic program, uses historical statistics and regression functions to estimate development costs, production costs, and the total program costs for the vehicle.

The executive program is configured to accept from the program user a particular sequence in which the constraints are to be satisfied. Using the initial parameters provided, it then establishes program loops and calls to the technology programs to specify completely an aerospace vehicle configuration, and executes this sequence. The data base provides the means of interchanging information between the technology programs. After a configuration is determined, it contains the information necessary to specify the configuration.

Optimization of the design enters after the loops are established to convert a list of requirements into a complete preliminary design. The final loop established by the executive program uses the design optimization program to perturb the initial parameters, and then determines a new configuration. Any improvement of the new configuration over the old configuration is determined, and the design optimization program is then used again to determine a new set of perturbations of the initial parameters. This process is continued until all requirements are satisfied, or no further improvements are determined.

The design optimization program is largely independent of the order in which the technology programs are called, and has available various search techniques. The search techniques available include sectioning search, creeping search, random point search, steepest descent search using numerically determined partial derivatives, quadratic search, Davidon search, pattern search, and random ray search. No one of these techniques is generally superior to the others.

The automated design procedure described above has been used for preliminary design of an advanced manned interceptor, and the optimization procedures were also used in a hypersonic transport design study [3]. In the hypersonic transport synthesis, the objective was to maximize passenger-carrying capacity over a 5500 nautical mile mission. With five parameters to vary, the nominal design produced 220 passengers. The five variable searches, independent of search technique, produced a payload of 253 passengers. Introducing five additional parameters raised the passenger count to 260. If, in addition, trajectory optimization was performed simultaneously, the number of passengers could be increased to 286 passengers.

B. Structural Design

The application of optimization techniques to automated design of structures will now be briefly reviewed. The most common mathematical model of structures used today for analysis and synthesis is the finite element model, which replaces the distributed structure by a lumped approximation composed of finite elements. The equilibrium equations for the elements then yield a large set of algebraic equations in the displacements, stresses, and design parameters. Due to the static indeterminancy of the structure, the design variables are not completely specified by these equations, but can be selected to satisfy some cost or performance function.

An optimization problem can thus be formulated [38, 39] where a cost, performance, or merit function $M(D)$ is to be minimized by varying the elements of the vector D of design parameters, subject to a set of constraints

$$h_j(D) \leq 0. \tag{95}$$

The object behind formulating this optimization problem is to obtain an automated method of sizing the elements; the basic arrangement of the elements is assumed to be given. Most commonly, the merit function is a minimum weight criterion for the structure. The set of constriants most frequently are formed from stress, strain, aeroelastic, and minimum gauge requirements. For a practical problem, the model might have 2000-5000 design parameters, 2000-4000 displacement degrees of freedom, and several different load conditions.

Two different approaches have been used to solve structural optimization problems [41, 42]. The first approach is a direct approach dealing directly with the merit function. In the direct method mathematical programming techniques such as gradient methods and direct search methods are used to vary the design parameters until the minimum of the merit function is found. The iterative process is usually started with a good, though nonoptimal, design as a nominal point. This procedure has the advantage that it can be used for general problems, and does not depend on special properties of the structure. However, it is limited because of computer speed and memory limitations to problems having only 200-300 design variables, which excludes it from application to many practical problems.

The second approach, or indirect approach, proceeds by deriving necessary conditions which an optimal structure must satisfy. Then, instead of trying to minimize directly the merit function, the necessary conditions are solved to determine the design parameter vectors of possible optimal structures. This procedure is analogous to the solution of calculus of variations problems by using the Euler-Lagrange equations, and the necessary conditions derived for the structure correspond to the Euler-Lagrange equations. The necessary conditions for an optimal structure are usually called optimality criteria.

For large problems, having several thousand design parameters, it has been found possible to derive optimality criteria resulting in equations which could be solved within the limits of available computer time and memory. The optimality criteria methods for large problems are simpler than the direct methods, and allow structures with several thousand design parameters to be designed optimally. This size problem (several thousand design parameters) is large enough to include many practical structural design problems, as for instance, wing box design. An example of an application to wing structure design is given in the work of Berke and Venkayya [41].

VI. SUMMARY

The preceding sections have shown that optimization techniques have been applied to a wide variety of aerospace vehicle analysis and synthesis problems. These applications have only been made posssible because of the recent availability of large digital computers, and limitations of the computers now available frequently determine if a particular optimization technique can be applied. Table I presents a synopsis of the state of application of the techniques. How to set up a meaningful optimization problem for the total aerospace vehicle is still not known; the preceding examples and problems have basically been

TABLE 1

SUMMARY AND STATUS OF OPTIMIZATION APPLICATIONS TO AEROSPACE VEHICLES

Factors	Concepts	Algorithms	Applications
Trajectory, Performance, Tactics			
1. Performance optimization	Developed	Off-line developed On-line under development	Presently in use
2. Air-to-air tactics	Being formulated	Being derived	Can presently only treat oversimplified problems
Tracking and Attitude Control			
1. Aircraft active control systems	Developed	Developed	The technology is presently being demonstrated
2. Pilot-vehicle analysis	Developed	Being developed	Data base must be accumulated before wide-spread application
3. MLS analysis	Developed	Developed	Applied to MLS analysis
Active landing gear	Developed	Developed	Awaiting technology demonstration
Reliability	Being formulated	Being derived	Awaiting concepts and algorithms
Vehicle and Structural Design			
1. Aircraft preliminary design	Being formulated	Being developed	Early versions in use
2. Structural design	Being formulated	Being developed	Awaiting more experience with the techniques

subsystem optimization, not system optimization. What quantity or quantities should be optimized in a "true" system optimization is uncertain; it seems likely that more than one optimality criteria may be required, and thus vector-valued cost functionals must be considered. Furthermore, many of the optimization techniques require that they be started near the optimal solution. They cannot handle large problem variations (as in aircraft configuration or structural configuration changes) and usually guarantee no more than a locally optimal solution. Nonetheless, "true" systems optimization is needed to make full use of modern flight vehicle concepts such as control configured vehicles. In short, subsystem integration is a very real current problem. How to accomplish this integration with modern optimization techniques is a very real challenge.

Despite this, optimization techniques have proved to be very valuable in aerospace vehicle design, as the variety of applications given in this chapter show. They have frequently provided methods and procedures to attack problems previously intractable, and systems designed using optimization techniques have often exhibited superior performance. Nevertheless, further work on optimization techniques for aerospace vehicle design is needed and should return significant benefits in the design of future systems.

REFERENCES

1. D.S. HAGUE, "Three-Degree-Of-Freedom Problem Optimization Formulation--Analytical Development", Part 1, Vol. III. AFFDL TR 64-1, Air Force Flight Dynamics Laboratory, Wright-Patterson AFB, Ohio, October 1964.

2. R.L. MOBLEY and R.R. VORWALD, "Three-Degree-Of-Freedom Problem Optimization Formulation--User's Manual", Part 2, Vol. III. AFFDL TR 64-1, Air Force Flight

Dynamics Laboratory, Wright-Patterson AFB, Ohio, October 1964.

3. D.S. HAGUE and C.R. GLATT, "Optimal Design Integration of Military Flight Vehicles". AFFDL TR 72-132, Air Force Flight Dynamics Laboratory, Wright-Patterson AFB, Ohio, December 1972.

4. S.K. LANDGRAF, AIAA J. 64, 288 (1964).

5. R. ISAACS, "Differential Games", Wiley, New York, 1965.

6. W.L. OTHLING, Jr., "Application of Differential Game Theory to Pursuit-Evasion Problems of Two Aircraft", Ph.D. Dissertation, Air Force Institute of Technology, Wright-Patterson AFB, Ohio, June 1970.

7. U.H.D. LYNCH, "Differential Game Barriers and Their Application in Air-to-Air Combat", Ph.D. Dissertation, Air Force Institute of Technology, Wright-Patterson AFB, Ohio, March 1973.

8. L.E. MILLER, Jr., "Application of Differential Games to Pursuit-Evasion Problems", Ph.D. Dissertation, Ohio State University, Columbus, Ohio, 1974.

9. A.L. LEATHAM, "Some Theoretical Aspects of Nonzero Sum Differential Games and Applications to Combat Problems", Ph.D. Dissertation, Air Force Institute of Technology, Wright-Patterson AFB, Ohio, June 1971.

10. G. STEIN and A.H. HENKE, "A Design Procedure And Handling Quality Criteria for Lateral-Directional Flight Control Systems". AFFDL TR 70-152, Air Force Flight Dynamics Laboratory, Wright-Patterson AFB, Ohio, May 1971.

11. A.J. VAN DIERENDONCK, C.R. STONE, and M.D. WARD, Proc. AIAA Guidance Control Conf. August 20-22, 1973, Key Biscayne, Fla.

12. R.P. DENARO, and G.L. GREENLEAF, "Selection of Optimal Stability Augmentation System Parameters for a High Performance Aircraft Using Pitch Paper Pilot".

Report No. GGC/EE/73-3, Air Force Institute of Technology, Wright-Patterson AFB, Ohio, October 1972.

13. Special Issue on Linear-Quadratic-Gaussian Problem, IEEE Trans. Automatic Control, AC-16, (6), (1971).

14. B. ETKIN, "Dynamics of Atmospheric Flight". Wiley, New York, 1972.

15. D. McRUER, I. ASHKENAS, and D. GRAHAM, "Aircraft Dynamics and Automatic Control". Princeton Univ. Press, Princeton, New Jersey, 1973.

16. E. WONG, "Stochastic Processes in Information and Dynamical Systems". McGraw-Hill, New York, 1971.

17. C.R. CHALK, T.P. NEAL, T.M. HARRIS, F.E. PRITCHARD, and R.J. WOODCOCK, "Background Information and User Guide for MIL-F-8785B(ASG), Military Specification--Flying Qualities of Piloted Airplanes". AFFDL TR 69-72, Air Force Flight Dynamics Laboratory, Wright-Patterson AFB, Ohio, August 1969.

18. R.A. SMITH, SIAM J. Appl. Math. 16, (1), 198-201 (1968).

19. H. KWAKERNAAK and R. SIVAN, "Linear Optimal Control Systems". Wiley, New York, 1972.

20. R.D. POYNEER, Symp. Air Force Appl. Modern Control Theory, Wright-Patterson AFB, July 9-11, 1974.

21. R.E. HEATH, III, "Optimal Incomplete Feedback Control of Linear Stochastic Systems", AFFDL TR 73-36, Air Force Flight Dynamics Laboratory, Wright-Patterson AFB, Ohio, June 1973.

22. S. AXSATER, Internat. J. Control, 4, (6), 549-566 (1966).

23. W.S. LEVINE and M. ATHANS, IEEE Trans. Automatic Control, AC-15 (1), 44-48 (1970).

24. W.M. WONHAM, SIAM J. Control, 6, 312 (1968).

25. D.G. LUENBERGER, IEEE Trans. Automatic Control, AC-16, (6) 596-602 (1971).

26. R.V. GRESSANG, Symp. Air Force Appl. Modern Control Theory,

Wright-Patterson AFB, July 9-11, 1974.

27. D. McRUER, and H. JEX, IEEE Trans. Human Factors Electronics, HFE-8 (3), 231-249 (1967).

28. D. McRUER and E. KRENDEL, "Mathematical Models of Human Pilot Behavior". AGARDOGRAPH No. 188, January 1974.

29. D.L. KLEINMAN, S. BARON, and W.H. LEVISON, IEEE Trans. Automatic Control, AC-16 (6) 824-832 (1971).

30. D.L. KLEINMAN, IEEE Trans. Automatic Control, AC-14, 524-527 (1964).

31. D.L. KLEINMAN and T.R. PERKINS, IEEE Trans. Automatic Control, AC-19 (4) 297-306 (1974).

32. T.R. HARVEY, "Application of an Optimal Control Pilot Model to Air-to-Air Combat". Report No. GA/MA/74 M-1, Air Force Institute of Technology, Wright-Patterson AFB, Ohio, March 1974.

33. R.O. ANDERSON, "A New Approach to the Specification and Evaluation of Flying Qualities", AFFDL TR 69-120, Air Force Flight Dynamics Laboratory, Wright-Patterson AFB, Ohio, June 1970.

34. J.D. ARNOLD, Proc. 14th JACC, Columbus, Ohio, June 1973, 800-803.

35. C.T. LEONDES and R.R. RANKINE, Jr., AIAA J. Aircraft, 8, (4) 286-293 (1972).

36. J.D. DILLOW, P.R. STOLZ, and M.D. ZUCKERMAN, "Analysis of Data Rate Requirements for Low Visibility Approach with a Scanning Beam Landing Guidance System". AFFDL TR 71-177, Air Force Flight Dynamics Laboratory, Wright-Patterson AFB, Ohio, February 1973.

37. R.R. HUBER, Jr. "Optimal Control Aircraft Landing Analysis" AFFDL TR 73-141, Air Force Flight Dynamics Laboratory, Wright-Patterson AFB, Ohio, December 1973.

38. C.D. CORSETTI, and J.D. DILLOW, "A Study of the Practibility of Active Vibration Isolation Applied to Aircraft

During the Taxi Conditions". AFFDL TR 71-159, Air Force Flight Dynamics Laboratory, Wright-Patterson AFB, Ohio, July 1972.

39. A.J. CONNORS, "Dynamic Programming for Optimum Redundancy in an Automatic Flight Control System". FDCC TM 63-1, Air Force Flight Dynamics Laboratory, Wright-Patterson AFB, Ohio, September 1963.

40. P. MAYBECK, "Failure Detection Through Functional Redundancy", AFFDL TR 7-03, Air Force Flight Dynamics Laboratory, Wright-Patterson AFB, Ohio, 1974.

41. L. BERKE and V.B. VENKAYYA, "Review of Optimality Criteria Approaches to Structural Optimization". Unpublished report, Air Force Flight Dynamics Laboratory, Wright-Patterson AFB, Ohio.

42. L. BERKE, and N.S. KHOT, "Use of Optimality Criteria Methods for Large Scale Systems", AGARD Lecture Series on Structural Optimization, October 10-18, 1974.

OPTIMIZATION OF DISTRIBUTED PARAMETER STRUCTURES UNDER DYNAMIC LOADS

E. J. HAUG

Concepts and Technology
AMSAR/RDT, U.S. Army Armaments Command
Rock Island, Illinois

T. T. FENG

Department of Mechanics
University of Iowa
Iowa City, Iowa

I. INTRODUCTION

Optimization of static structural systems has been treated extensively in the literature. Recent work [1] has resulted in a practical method for treating dynamic finite-dimensional structures. There are, however, many problems in mechanical system design in which one or more space variables and a time variable must be treated. This is the case for structural and machine design problems involving transient dynamic response and continuous spatial distribution of material. Such problems involve partial differential equations and are commonly called "distributed parameter problems".

A relatively extensive literature on distributed parameter control systems has developed since the mid 1960's. It ranges from the very rigorous treatment of a restricted class of problems [2] to practical and in-depth treatments of special classes of thermal and hydrodynamics problems [3]. An extensive collection of distributed control applications is also available [4].

In this chapter, a state space steepest descent technique is developed and applied to distributed parameter structural dynamic optimization problems. The approach presented herein represents an extension of the steepest descent method for static structural problems [5] to the dynamic response domain.

II. DISTRIBUTED PARAMETER DYNAMIC SYSTEM OPTIMIZATION

A major class of structural optimization problems involves spatial distribution of a design variable and transient dynamic performance constraints. Performance of these systems is described by a family of hyperbolic partial differential equations, sometimes called "evolution equations" [6].

A. Problem Formulation

The domain of the independent variable will be taken as a product space $\Omega \times T$, where $\Omega \subset R^K$ and $T = [0, \tau]$. The spatial boundary of Ω is denoted as Γ, so the lateral boundary of the set $\Omega \times T$ is $S = \Gamma \times T$, as shown in Fig. 1. Here, the scalar time dimension is denoted by t and the vector space dimension is denoted by a vector x.

To distinguish the class of mechanical design problems from feedback control problems, the design variables considered here are only space dependent. That is, $u(x)$ is defined on Ω for all t. The variable b is a vector of scalar design parameters. These design variables and parameters are to be chosen when the system is constructed and do not vary with the temporal variable t.

The state of the system, generally a displacement field, is denoted by $z(x, t)$ and is both space and time dependent. The dynamic system problems treated here are described by initial boundary value problems that are linear in the state variable and can be written in differential operator notation as

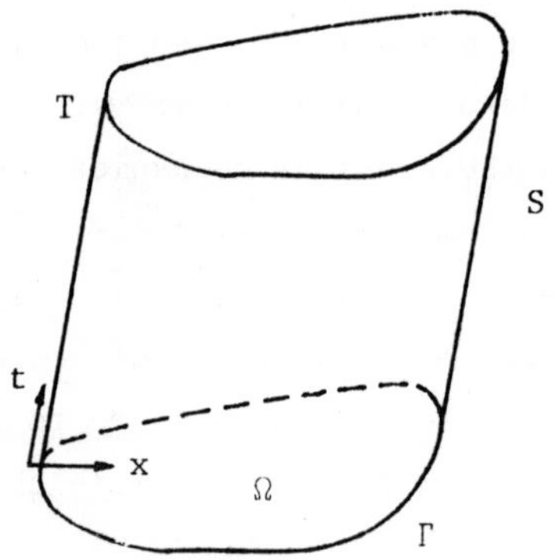

Fig. 1. Domain of the Independent Variables

$$L(u, b)z + M(u, b)z = Q(u, b, x, t) \quad \text{in} \quad \Omega \times T, \tag{1}$$

with boundary conditions

$$Bz = q(x, t) \qquad \text{on} \quad S = \Gamma \times T \tag{2}$$

and initial conditions

$$Cz = r(x) \quad \text{in} \quad \Omega \quad \text{with} \quad t = 0. \tag{3}$$

In Eq. (1), the linear differential operators L and M are spatial and temporal operators, respectively. The coefficients in these operators may depend on the design variables $u(x)$ and parameters b. The operator L will be presumed symmetric for z satisfying homogeneous boundary conditions in Eq. (2) (i.e., with $q = 0$).

The cost functional and constraints are taken in the form

$$\psi_0 = g_0(b) + \int_0^\tau \int_\Gamma h_0(z, b)\, ds\, dt + \int_0^\tau \int_\Omega f_0(z, u, b)\, dx\, dt, \tag{4}$$

$$\psi_\alpha = g_\alpha(b) + \int_0^\tau \int_\Gamma h_\alpha(z, b)\, ds\, dt + \int_0^\tau \int_\Omega f_\alpha(z, u, b)\, dx\, dt$$

$$\begin{cases} = 0, & \alpha = 1, \ldots, r', \\ \leq 0, & \alpha = r' + 1, \ldots, r \end{cases} \tag{5}$$

and

$$\phi_\beta(x, u) \begin{cases} = 0, & \beta = 1, \ldots, q', \\ \leq 0, & \beta = q' + 1, \ldots, q, \end{cases} \qquad x \in \Omega. \tag{6}$$

The arguments of all functions appearing in Eqs. (1)-(6) may depend on x and t explicitly. These variables are suppressed here for notational convenience. It may also be noted that the state variable does not appear in the pointwise constraints of Eq. (6). It is presumed that pointwise constraints of the form

$$\eta(x,\ t,\ u,\ z,\ b) \leq 0 \qquad \text{in} \quad \Omega \times T \tag{7}$$

have been replaced by an equivalent functional constraint

$$\int_0^\tau \int_\Omega [\eta + |\eta|]\ dx\ dt = 0. \tag{8}$$

Finally, it may be noted that functionals involving integration only over Ω or Γ can be transformed to the form of one of the two multiple integrals of Eqs. (4) and (5) by multiplying by $1/\tau$ and integrating from zero to τ. No loss in computational efficiency results from this transformation.

B. Design Sensitivity Analysis

The problem formulation presented here is especially well suited to dynamic system optimal design. Special features of the problem can be exploited to obtain an effective computational algorithm. Prior to developing a steepest descent programming algorithm for this dynamic system problem, it is necessary to determine the effect of a perturbation in design $(\delta u,\ \delta b)$ on the functionals of Eqs. (4) and (5). For definiteness, the cost functional will be considered. Results of this analysis will carry over to the constraint functionals. Sensitivity analysis of the pointwise constraints of Eq. (6) is very simple and may be summarized in algebraic form as

$$\delta\phi = \frac{\partial\phi}{\partial u}\,\delta u\,. \tag{9}$$

The linearized form for perturbation of ψ_0 is

$$\delta\psi_0 = \frac{\partial g_0}{\partial b}\,\delta b + \int_0^\tau\!\!\int_\Gamma \left[\frac{\partial h_0}{\partial z}\,\delta z + \frac{\partial h_0}{\partial b}\,\delta b \right] ds\; dt$$

$$+ \int_0^\tau\!\!\int_\Omega \left[\frac{\partial h_0}{\partial z}\,\delta z + \frac{\partial f_0}{\partial u}\,\delta u + \frac{\partial f_0}{\partial b}\,\delta b \right] dx\; dt. \tag{10}$$

The next step in developing design sensitivity data is elimination of δz from Eq. (10), through use of adjoint equations associated with Eqs. (1)-(3).

First, the linearized initial boundary value problem of Eqs. (1)-(3) is

$$L\;\delta z + \Delta_u(Lz) + \frac{\partial(Lz)}{\partial b}\,\delta b + M\delta z + \Delta_u(Mz) + \frac{\partial(Mz)}{\partial b}\,\delta b$$

$$= \frac{\partial Q}{\partial u}\,\delta u + \frac{\partial Q}{\partial b}\,\delta b \qquad \text{in} \quad \Omega \times T, \tag{11}$$

$$B\;\delta z = 0 \qquad \text{on} \quad S, \tag{12}$$

and

$$C\;\delta z = 0 \qquad \text{in} \quad \Omega, \quad t = 0, \tag{13}$$

where Δ_u denotes variation of a quantity with respect to only a change in u.

In defining the adjoint equations to be used in eliminating δz from Eq. (10), the following identity is required:

$$\int_0^\tau \int_\Omega \lambda^T [L\ \delta z + M\ \delta z]\ dx\ dt = \int_0^\tau [\int_\Omega \lambda^T L\ \delta z\ dx]\ dt + \int_\Omega [\int^\tau \lambda^T M\ \delta z\ dt]\ dx, \quad (14)$$

where iterated integrals are employed in Eq. (14). Integration by parts over Ω in the first spatial integral on the right of Eq. (14) yields

$$\int_\Omega \lambda^T L\ \delta z\ dx = \int_\Omega \delta z^T L\lambda\ dx + \int_\Gamma A(\lambda,\ \delta z)\ ds \quad \text{on}\ T, \quad (15)$$

where λ is called an "adjoint variable" and $A(\lambda,\ \delta z)$ is a bilinear form in λ, δz, and their spatial derivatives. Since the operator L is symmetric, one can cause the boundary integral in Eq. (15) to vanish by choosing $B\lambda = 0$. For present purposes, however, one chooses the boundary condition on λ such that the boundary integral in Eq. (15) is related to the term $\int_\Gamma \frac{\partial h_0}{\partial z}\ \delta z\ ds$ that arises in Eq. (10).

Similarly, the temporal integral in the second term of Eq. (14) may be integrated by parts to eliminate all time derivatives of δz and obtain

$$\int_0^\tau \lambda^T M\ \delta z\ dt = \int_0^\tau \delta z^T M\lambda\ dt + H(\lambda,\ \delta z)\Big|_0^\tau \quad \text{in}\ \Omega, \quad (16)$$

where the bilinear form $H(\lambda,\ \delta z)$ involves temporal derivatives of λ and δz. One can cause this bilinear form to vanish by requiring that

$$C\lambda = 0 \quad \text{in}\ \Omega, \quad t = \tau. \quad (17)$$

One may now substitute Eqs. (15) and (16) into Eq. (14) and change orders of integration to obtain

$$\int_0^\tau \int_\Omega \lambda^T [L\ \delta z + M\ \delta z]\ dx\ dt = \int_0^\tau \int_\Omega \delta z^T [L\lambda + M\lambda]\ dx\ dt$$

$$+ \int_0^\tau \int_\Gamma A(\lambda,\ \delta z)\ ds\ dt. \tag{18}$$

To utilize this identity, choose λ to satisfy the adjoint differential equation

$$L\lambda + M\lambda = \frac{\partial f_0^T}{\partial z} \qquad \text{in} \quad \Omega \times T \tag{19}$$

and choose boundary and terminal conditions on λ such that

$$\int_0^\tau \int_\Gamma \frac{\partial h_0}{\partial z}\ \delta z\ ds\ dt = \int_0^\tau \int_\Gamma A(\lambda,\ \delta z)\ ds\ dt$$

holds identically for δz satisfying Eqs. (12) and (13). In view of the symmetry of the operators L and M, the resulting boundary conditions on λ are of the form

$$B\lambda = \bar{q}(x,\ t) \qquad \text{on} \quad \Gamma \times T \tag{20}$$

where $\bar{q}(x,\ t)$ depends on the form of the function $h_0(z,\ b)$.

Substituting from Eqs. (19) and (20) into the right side of Eq. (18) and for $L\ \delta z + M\ \delta z$ from Eq. (11) into the left side of Eq. (18), one obtains

$$\int_0^\tau \int_\Gamma \frac{\partial h_0}{\partial z} \delta z \, ds \, dt + \int_0^\tau \int_\Omega \frac{\partial f_0}{\partial z} \delta z \, dx \, dt$$

$$= \int_0^\tau \int_\Omega \lambda^T \left[\frac{\partial Q}{\partial u} \delta u + \frac{\partial Q}{\partial b} \delta b - \Delta_u (Lz)\right.$$

$$\left. - \frac{\partial (Lz)}{\partial b} \delta b - \Delta_u (Mz) - \frac{\partial (Mz)}{\partial b} \delta b\right] dx \, dt. \tag{21}$$

Equation (21) now provides an identity that can be used to eliminate dependence on δz in Eq. (10). Making this substitution, one obtains

$$\delta\psi_0 = \ell^{\psi_0^T} \delta b + \int_\Omega \Lambda^{\psi_0^T} \delta u \, dx, \tag{22}$$

where

$$\ell^{\psi_0} = \frac{\partial g_0^T}{\partial b} + \int_0^\tau \int_\Gamma \frac{\partial h_0^T}{\partial b} ds \, dt$$

$$+ \int_0^\tau \int_\Omega \left[\frac{\partial f_0}{\partial b} + \frac{\partial Q^T}{\partial b} \lambda - \frac{\partial (Lz)^T}{\partial b} \lambda - \frac{\partial (Mz)^T}{\partial b} \lambda\right] dx \, dt \tag{23}$$

$$\Lambda^{\psi_0}(x) = \int_0^\tau \left[\frac{\partial f_0^T}{\partial u} + \frac{\partial Q^T}{\partial u} \lambda - \hat{L}(z, \lambda) - \hat{M}(z, \lambda)\right] dt, \tag{24}$$

and the terms $\hat{L}(z, \lambda)$ and $\hat{M}(z, \lambda)$ are bilinear forms in z and λ that are obtained as coefficients of δu in Eq. (21), by integrating $\lambda^T \Delta u(Lz)$ and $\lambda^T \Delta u(Mz)$ by parts to eliminate any derivatives of δu that might occur.

Making completely analogous computations and definitions associated with the functional ψ_α, i.e., developing and solving adjoing initial boundary value problems, using the functions and integrands of Eq. (5) instead of Eq. (4), one obtains an analogous sensitivity formula

$$\delta\psi_\alpha = \ell^{\psi_\alpha^T} \delta b + \int_\Omega \Lambda^{\psi_\alpha^T} \delta u \, dx, \qquad \alpha = 1, \ldots, r. \tag{25}$$

One now has sensitivity data that is independent of time. That is, the sensitivity data is given in the design variable space. One might view the process of eliminating this time dependence as collapsing the time variable out of the sensitivity analysis problem.

III. A STEEPEST DESCENT COMPUTATIONAL ALGORITHM

A form of steepest descent or gradient projection method may be readily developed using the total design differentials or sensitivity coefficients that have been calculated in design space. Such techniques have been used effectively for control and design problems with one independent variable [7]. An extension of the technique for distributed parameter structural optimization has recently been presented by Haug and co-workers [5].

One wishes to select a change in design $(\delta b, \delta u)$ that decreases $\delta\psi_0$ of Eq. (22) as much as possible, while satisfying constraints

$$\delta\tilde{\psi}_\alpha = \ell^{\tilde{\psi}_\alpha^T} \delta b + \int_\Omega \Lambda^{\tilde{\psi}_\alpha} \delta u \, d\Omega$$

$$\begin{cases} = \Delta\psi_\alpha, & \alpha = 1, \ldots, r' \\ \leq \Delta\tilde{\psi}_\alpha, & \alpha = r' + 1, \ldots, r \quad \text{and} \quad \tilde{\psi}_\alpha \geq 0 \end{cases} \tag{26}$$

for all α such that $\tilde{\psi}_\alpha \geq 0$,

$$\delta\tilde{\phi}_\beta = \frac{\delta\tilde{\phi}_\beta}{\partial u}\,\delta u \leq \Delta\tilde{\phi}_\beta \qquad \text{in } \Omega, \tag{27}$$

for all β and $x \in \Omega$ such that $\tilde{\phi}_\beta(\ , u(x)) \geq 0$, and

$$\delta b^T W_b\, \delta b + \int_\Omega \delta u^T\, W_u\, \delta u\, d\Omega \leq \xi^2. \tag{28}$$

Here, a tilde denotes constraints that are violated or tight and must be enforced. In Eq. (26), $\Delta\tilde{\psi}_\alpha$ is the correction in constraint error desired, normally $\Delta\tilde{\psi}_\alpha = -\tilde{\psi}_\alpha$. Similarly, $\Delta\tilde{\phi}_\beta$ is a correction in pointwise constraint error, normally $\Delta\tilde{\phi}_\beta = -\tilde{\phi}_\beta$. The constraint of Eq. (28) is simply a form of step size restrictions to assure that $(\delta b, \delta u)$ is small enough so that the linear approximations used in computing design sensitivity coefficients are sufficiently accurate. Finally, W_b and W_u are simply positive definite weighting matrices and ξ is a small parameter.

The necessary conditions of the calculus of variations may now be applied directly to find $(\delta b, \delta u)$ to minimize $\delta\psi_0$, subject to constraints of Eqs. (26)-(28). There exist multipliers γ_α, $\gamma_\alpha \geq 0$ for $\alpha > r'$, such that $\psi_\alpha \geq 0$; $\mu_\beta(x)$, $\mu_\beta(x) \geq 0$ for $\beta > q'$, such that $\phi_\beta(x) \geq 0$; and $\nu \geq 0$ for which the following conditions must hold:

$$\ell^{\psi_0} + \ell^{\tilde{\psi}^T}\,\tilde{\gamma} + 2\nu W_b\,\delta b = 0, \tag{29}$$

$$\Lambda^{\psi_0} + \Lambda^{\tilde{\psi}^T}\,\tilde{\gamma} + \frac{\partial\tilde{\phi}^T}{\partial u}\,\mu + 2\nu W_u\,\delta u = 0 \qquad \text{in } \Omega, \tag{30}$$

$$\tilde{\gamma}_{\alpha}(\delta\tilde{\psi}_{\alpha} - \Delta\tilde{\psi}_{\alpha}) = 0, \tag{31}$$

$$\tilde{\mu}_{\beta}(x)(\delta\tilde{\phi}_{\beta} - \Delta\tilde{\phi}_{\beta}) = 0 \qquad \text{in } \Omega, \tag{32}$$

and

$$\nu[\delta b^T W_b \delta b + \int_{\Omega} \delta u^T W_u \delta u \, d\Omega - \xi^2] = 0. \tag{33}$$

Following a successfully applied technique for similar problems [5], it is assumed that tight constraints will continue to be tight, so the coefficients of $\tilde{\gamma}_{\alpha}$, $\tilde{\mu}_{\beta}(x)$, and ν in Eqs. (31)-(33) will be presumed zero. The remaining equations will then be solved and the algebraic signs of the multipliers checked to see if the assumption is valid. If negative multipliers corresponding to inequality constraints arise, the corresponding constraints are relaxed.

Solving Eqs. (29) and (30) for δb and δu, one obtains

$$\delta b = -\frac{1}{2\nu} W_b^{-1} (\ell^{\psi_0} + \ell^{\tilde{\psi}^T} \tilde{\gamma}) \tag{34}$$

and

$$\delta u = -\frac{1}{2\nu} W_u^{-1} (\Lambda^{\psi_0} + \Lambda^{\tilde{\psi}^T} \tilde{\gamma} + \frac{\partial\tilde{\phi}}{\partial u} \tilde{\mu}). \tag{35}$$

Substituting these expressions into Eqs. (26) and (27), both taken as equalities, one may solve for $\tilde{\gamma}$ and $\tilde{\mu}$ as [5]:

$$\tilde{\gamma} = -M_{\psi\psi}^{-1}[2\nu(\Delta\tilde{\psi} - M_{\psi\phi}) + M_{\psi\psi_0}] \tag{36}$$

and

$$\tilde{\mu}(x) = -(\frac{\partial\tilde{\phi}}{\partial u} W_u^{-1} \frac{\partial\tilde{\phi}^T}{\partial u})^{-1} [2\nu \, \Delta\phi + \frac{\partial\tilde{\phi}}{\partial u} W_u^{-1} (\Lambda^{\psi_0} + \Lambda^{\tilde{\psi}^T} \tilde{\gamma})], \tag{37}$$

where

$$M_{\psi\psi_0} = \int_\Omega \Lambda^{\tilde{\psi}^T} W_u^{-1} [I - \frac{\partial\tilde{\phi}^T}{\partial u} (\frac{\partial\tilde{\phi}}{\partial u} W_u^{-1} \frac{\partial\tilde{\phi}^T}{\partial u})^{-1} \frac{\partial\tilde{\phi}}{\partial u} W_u^{-1}] \Lambda^{\psi_0} \, d\Omega$$

$$+ \ell^{\tilde{\psi}^T} W_b^{-1} \ell^{\psi_0}, \tag{38}$$

$$M_{\psi\psi} = \int_\Omega \Lambda^{\tilde{\psi}^T} W_u^{-1} [I - \frac{\partial\tilde{\phi}^T}{\partial u} (\frac{\partial\tilde{\phi}}{\partial u} W_u^{-1} \frac{\partial\tilde{\phi}^T}{\partial u})^{-1} \frac{\partial\tilde{\phi}}{\partial u} W_u^{-1}] \Lambda^{\tilde{\psi}} \, d\Omega$$

$$+ \ell^{\tilde{\psi}^T} W_b^{-1} \ell^{\tilde{\psi}}, \tag{39}$$

and

$$M_{\psi\phi} = \int_\Omega \Lambda^{\tilde{\psi}^T} W_u^{-1} \frac{\partial\tilde{\phi}^T}{\partial u} (\frac{\partial\tilde{\phi}}{\partial u} W_u^{-1} \frac{\partial\tilde{\phi}^T}{\partial u})^{-1} \Delta\phi \, d\Omega. \tag{40}$$

Substituting from Eqs. (36) and (37) into Eqs. (34) and (35), one may obtain a reduced formula for δb and δu in the form

$$\delta b = -\frac{1}{2\nu} \delta b^1 + \delta b^2 \tag{41}$$

and

$$\delta u(x) = -\frac{1}{2\nu} \delta u^1(x) + \delta u^2(x), \tag{42}$$

where

$$\delta b^1 = W_b^{-1} [\ell^{\psi_0} - \ell^{\tilde{\psi}} M_{\psi\psi}^{-1} M_{\psi\psi_0}], \tag{43}$$

$$\delta b^2 = W_b^{-1} \ell^{\tilde{\psi}} M^{-1} [\Delta\tilde{\psi} - M_{\psi\phi}], \tag{44}$$

$$\delta u^1(x) = W_u^{-1}[I - \frac{\partial\tilde{\phi}^T}{\partial u}(\frac{\partial\tilde{\phi}}{\partial u} W_u^{-1} \frac{\partial\tilde{\phi}^T}{\partial u})^{-1} \frac{\partial\tilde{\phi}}{\partial u} W_u^{-1}]$$

$$\times [\Lambda^{\psi_0} - \Lambda^{\tilde{\psi}} M_{\psi\psi}^{-1} M_{\psi\psi_0}], \qquad x \in \Omega, \tag{45}$$

$$\delta u^2(x) = W_u^{-1}[I - \frac{\partial\tilde{\phi}^T}{\partial u}(\frac{\partial\tilde{\phi}}{\partial u} W_u^{-1} \frac{\partial\tilde{\phi}^T}{\partial u})^{-1} \frac{\partial\tilde{\phi}}{\partial u} W_u^{-1}]$$

$$\times [\Lambda^{\tilde{\psi}} M_{\psi\psi}^{-1} (\Lambda\tilde{\psi} - M_{\psi\phi})]$$

$$+ W_u^{-1} \frac{\partial\tilde{\phi}^T}{\partial u}(\frac{\partial\tilde{\phi}}{\partial u} W_u^{-1} \frac{\partial\tilde{\phi}^T}{\partial u})^{-1} \Delta\tilde{\phi}, \qquad x \in \Omega. \tag{46}$$

While these formulas appear to be somewhat complex, they are readily computable. Further, they can be routinely programmed for digital computation. It can also be shown that $(\delta b^1, \delta u^1)$ and $(\delta b^2, \delta u^2)$ are orthogonal [8] and that a necessary condition for convergence to the solution is that δb^1 and $\delta u^1(x)$ must approach zero [8]. This condition provides a good check for convergence.

If the constraints are all satisfied, $\delta u^2(x)$ and δb^2 will all be zero. In this case,

$$\Delta\psi_0 = -\frac{1}{2\nu}[M_{\psi_0\psi_0} - M_{\psi\psi_0}^T M_{\psi\psi}^{-1} M_{\psi\psi_0}], \tag{47}$$

where

$$M_{\psi_0\psi_0} = \ell^{\psi_0^T} W_b \ell^{\psi_0} + \int_\Omega \Lambda^{\psi_0^T} W_u^{-1}$$

$$\times (I - \frac{\partial\tilde{\phi}^T}{\partial u}[\frac{\partial\tilde{\phi}}{\partial u} W_u^{-1} \frac{\partial\tilde{\phi}^T}{\partial u}]^{-1}$$

$$\times \frac{\partial\tilde{\phi}}{\partial u} W_u) \Lambda^{\psi_0} \, d\Omega. \tag{48}$$

One can now specify a reasonable, desired reduction $\Delta\psi_0$ in ψ_0 and determine the parameter ν from Eq. (47) to yield the desired reduction in the cost function.

In summary form, one has the following steepest descent algorithm:

Step 1. Make an engineering estimate $u^{(0)}(x)$ and $b^{(0)}$ of the optimum design functions and parameter.

Step 2. Solve Eqs. (1)-(3) for $z^{(0)}(x, t)$ corresponding to $u^{(0)}(x)$ and $b^{(0)}$.

Step 3. Check constraints and form vectors of tight and violated constraints $\tilde{\psi}$ and $\tilde{\phi}$, of Eqs. (5) and (6).

Step 4. Solve the differential equation (19) subject to the boundary conditions of Eq. (20) and initial conditions of Eq. (17); corresponding to both the functionals ψ_0 and $\tilde{\psi}_\alpha$; for λ^{ψ_0} and λ^{ψ_α}, respectively.

Step 5. Compute $\Lambda^{\psi_0}(x)$, ℓ^{ψ_0}, $\Lambda^{\tilde{\psi}}(x)$, and ℓ^{ψ} in Eqs. (23) and (24).

Step 6. Choose constraint corrections $\Delta\tilde{\psi}_\alpha$ and $\Delta\tilde{\phi}_\beta$.

Step 7. Compute $M_{\psi\psi_0}$, $M_{\psi\psi}$, and $M_{\psi\phi}$ in Eqs. (38)-(40).

Step 8. Choose $\nu > 0$ and compute $\tilde{\gamma}$ and $\tilde{\mu}(x)$ in Eqs. (36) and (37). If any components of $\tilde{\gamma}$ with $\alpha > r'$ or $\tilde{\mu}(x)$ with $i > q'$ are negative, redefine $\tilde{\psi}$ or $\tilde{\phi}(x)$ by deleting corresponding terms and return to Step 5.

Step 9. Compute $\delta u^1(x)$, $\delta u^2(x)$, δb^1, and δb^2 in Eqs. (43)-(46).

Step 10. Compute

$$u^{(1)}(x) = u^{(0)}(x) - \frac{1}{2\nu}\,\delta u^1(\ ) + \delta\ ^2(\ )$$

and

$$b^{(1)} = b^{(0)} - \frac{1}{2\nu}\,\delta b^1 - \delta b^2.$$

Step 11. If the constraints are satisfied and $\|\delta u^1(x)\|$ and $\|\delta b^1\|$ are sufficiently small, terminate, otherwise, return to Step 2 with $u^{(0)}(x)$ and $b^{(0)}$ replaced by $u^{(1)}(x)$ and $b^{(1)}$, respectively.

It may be noted that if there is no spatial variable in the problems and the state variable reduces to a time-dependent vector, as is the case when one treats a finite-element structure, the optimization problem is considerably simplified. In this case, there is only a design parameter and the spatial operator $L(b)$ is just a matrix. The above algorithm applies and considerable efficiencies can be effected by taking advantage of the simplified mathematical structure. An in-depth treatment of this finite-dimensional dynamics problem has been carried out, with good success [1]. The reader is referred to this literature for the reduced problem. Attention is restricted in the following examples to distributed design over one (beam) and two (plate) [9] space dimensions.

IV. DYNAMIC OPTIMIZATION OF BEAMS

As a first example, one wishes to distribute material along a beam (one space dimension) to minimize weight, subject to constraints on transient dynamic response. For simplicity, beams treated herein are loaded by uniform spatially distributed, time-varying loads. For numerical implementation of the steepest descent algorithm, a finite-element analysis method is employed [10] to obtain approximate solutions of the equations of motion. Subspace iteration [11] is used to solve the eigenvalue problem and modal analysis is used to solve the equations of motion and the adjoint equations. The computer used was an IBM 360/65.

A. Example 1: A Simply Supported Beam

For the simply supported beam of Fig. 2, one wishes to minimize total weight, subject to the condition that displacement is always within specified bounds and that the cross-sectional area $u(x)$ and geometrically similar cross sections are employed, so $I = \alpha u^2$ [5]. The system dynamic equations are

$$[E\alpha u^2 z'']'' + \rho u\ddot{z} = Q(x, t) \quad \text{in } \Omega \times T, \tag{49a}$$

where $\frac{\partial z}{\partial x} \equiv z'$ and $\frac{\partial z}{\partial t} \equiv \dot{z}$, with boundary conditions

$$z(0, t) = \frac{\partial^2 z}{\partial x^2}(0, t) = 0 \quad \text{in } T, \tag{49b}$$

$$z(a, t) = \frac{\partial^2 z}{\partial x^2}(a, t) = 0 \quad \text{in } T \tag{49c}$$

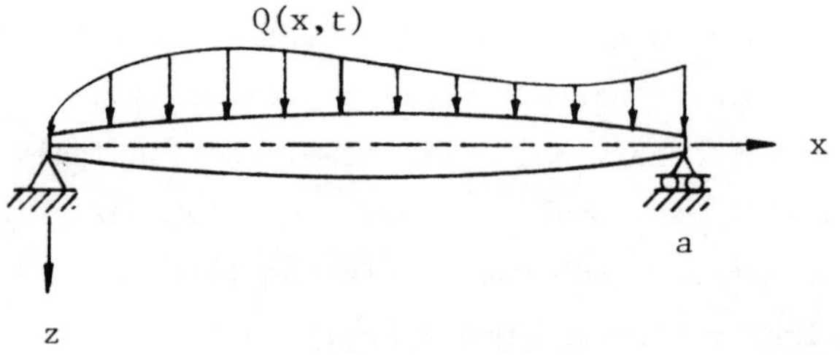

Fig. 2. Simply Supported Beam with Dynamic Load

and initial conditions

$$z(x, 0) = \frac{\partial z}{\partial t}(x, 0) = 0 \qquad \text{in} \quad \Omega. \tag{49d}$$

In this problem, the space domain is $\Omega = [0, a]$ and Γ is just the two points $x = 0$ and a.

One may readily verify that the operators $Lz = (E\alpha u^2 z'')''$ and $Mz = \rho u\ddot{z}$ are symmetric in the sense defined in Section II. Further, $\Delta_u(Lz) = (2E\alpha u\ \delta u\ z'')''$ and $\Delta_u(Mz) = \rho\ \delta u\ \ddot{z}$. Simple integrations by parts in Eq. (21) yield

$$\begin{aligned}\int_\Omega \lambda\Delta_u(Lz)\, dx &= \int_\Omega \lambda(2E\alpha u\ \delta u\ z'')''\, dx \\ &= \int_\Omega [2E\alpha u\lambda''\ z'']\ \delta u\, dx \equiv \int_\Omega \hat{L}(z,\lambda)\ \delta u\, dx\end{aligned} \tag{50}$$

and

$$\begin{aligned}\int_\Omega\int_0^\tau \lambda\Delta_u(Mz)\, dx\, dt &= \int_\Omega\int_0^\tau [\lambda\rho\ddot{z}]\ \delta u\, dx\, dt \\ &= \int_\Omega\int_0^\tau [-\rho\dot{\lambda}\dot{z}]\ \delta u\, dx\, dt \\ &\equiv \int_\Omega\int_0^\tau \hat{M}(z, \lambda)\ \delta u\, dx\, dt.\end{aligned} \tag{51}$$

The weight to be minimized is

$$\psi_0 = \int_0^a \gamma u\, dx \quad \text{or} \quad \psi_0 = \int_0^\tau\int_0^a \frac{\gamma}{\tau}\, u(x)\, dx\, dt, \tag{52}$$

where γ is material weight density. The displacement constraint

$$|z(x, t)| \leq d \qquad \text{in} \quad \Omega \times T \tag{53}$$

is transformed to functional form as

$$\psi_1 = \int_0^\tau \int_0^a [|z| - d + \Big||z| - d\Big|]\, dx\, dt = 0. \tag{54}$$

Finally, a lower bound is placed on the cross-sectional area. This is

$$\phi_1 = -u(x) + u_0 \leq 0 \qquad \text{in } \Omega, \tag{55}$$

where $u_0 > 0$ is given.

The functional constraint of Eq. (54) requires solution of an adjoint problem; namely, Eq. (19). By inspection, this is just Eq. (49), with the applied load replaced by

$$\bar{Q} = \frac{\partial f_1}{\partial z} = \text{sgn}(z)[1 + \text{sgn}(|z| - d)] \tag{56}$$

and with initial conditions replaced by terminal conditions

$$\lambda(x, t) = \dot{\lambda}(x, \tau) = 0 \qquad \text{in } \Omega. \tag{57}$$

Again, by inspection $\Lambda^{\psi_0} = \gamma$ and $\dfrac{\partial \phi_1}{\partial u} = -1$. From Eqs. (24), (50), (51), and (54),

$$\Lambda^{\psi_1}(x) = \int_0^\tau [-2E\alpha u\lambda'' z'' + \rho\dot{\lambda}\dot{z}]\, dt. \tag{58}$$

All information is now available to implement the optimization algorithm of Section III.

The simply supported beam and its load are shown in Fig. 3. Constraint parameters are $d = 0.4$ in and $u_0 = 0.2513$ in^2. Due to symmetry, only half of the beam is analyzed numerically. The half beam is divided into 10 finite elements, with a total

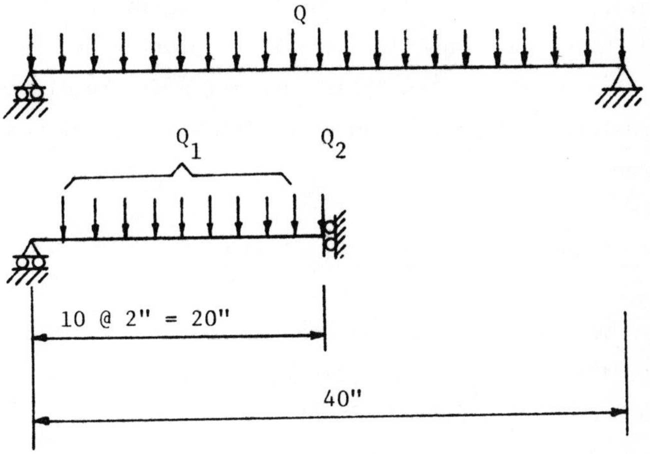

$\alpha = 0.3$, $E = 30 \times 10^6$ psi, $\gamma = 0.28$ lb/in^3

$\dot{z}(x,0) = z(x,0) = 0$

$$Q = \begin{cases} 9 \sin(80\pi t) \text{ lb/in}, & 0 \leq t \leq 0.0125 \text{ sec} \\ 0 \text{ lb/in}, & 0.0125 \leq t \leq 0.0150 \text{ sec} \end{cases}$$

$$Q_1 = \begin{cases} 18 \sin(80\pi t) \text{ lb}, & 0 \leq t \leq 0.0125 \text{ sec} \\ 0 \text{ lb}, & 0.0125 \leq t \leq 0.0150 \text{ sec} \end{cases}$$

$$Q_2 = \begin{cases} 9 \sin(80\pi t) \text{ lb}, & 0 \leq t \leq 0.0125 \text{ sec} \\ 0 \text{ lb}, & 0.0125 \leq t \leq 0.0150 \text{ sec} \end{cases}$$

Fig. 3. Simply supported beam subjected to uniformly distributed load with $\alpha = 0.3$, $E = 30 \times 10^6$ psi, $\gamma = 0.28$ lb/in.^{3}, $\dot{z}(x, 0) = z(x, 0) = 0$, and with values Q, Q_1, and Q_2 as shown.

of 20 degrees of freedom. Five eigenvectors were used for modal analysis. The initial design shown in Table 1 weighed 2.184 lb and $\|\delta u^1\| = 1.252$. The displacement constraint was violated by 4.5% in the 6th iteration and remained active until the 22nd iteration, where $\|\delta u^1\| = 0.0559$. Results are shown in Table 1 and the weight reduction history is plotted in Fig. 8 of Example 3. Numerical results obtained using a finite-dimensional formulation [1] are also given in Table 1, for comparative purposes.

B. Example 2: A Fixed-Fixed Beam

To further illustrate the method, the fixed-fixed beam of Fig. 4 is treated. The only modification required is to change the boundary conditions in Eq. (49) to

$$\left.\begin{aligned} z(0, t) &= \frac{\partial z}{\partial x}(0, t) = 0 \\ z(a, t) &= \frac{\partial z}{\partial x}(a, t) = 0 \end{aligned}\right\} \text{ in } T. \tag{59}$$

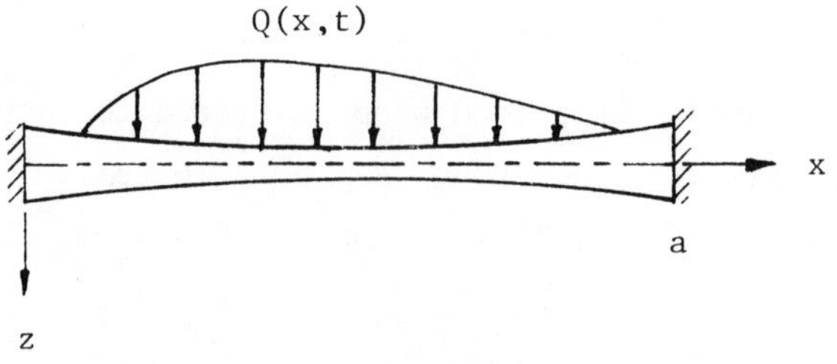

Fig. 4. Fixed-Fixed Beam with Dynamic Load

TABLE 1

INITIAL AND FINAL DESIGNS OF EXAMPLES 1, 2, AND 3

Examples:	1. Simply Supported Beam			2. Clamped Beam			3. Cantilever Beam		
Designs (in^2)	Initial	Final	Discrete[a]	Initial	Final	Discrete[a]	Initial	Final	Discrete[a]
u_1	0.3000	0.2513	0.2513	0.4000	0.3511	0.3351	0.4800	0.3183	0.3143
u_2	0.3200	0.2513	0.2513	0.3800	0.3471	0.3331	0.4600	0.2975	0.2974
u_3	0.3400	0.2513	0.2513	0.3600	0.3045	0.3080	0.4400	0.2848	0.2812
u_4	0.3600	0.2808	0.2513	0.3400	0.2513	0.2513	0.4200	0.2683	0.2861
u_5	0.3800	0.3313	0.3702	0.3200	0.2513	0.2513	0.4000	0.2516	0.2513
u_6	0.4000	0.3795	0.3975	0.3000	0.2513	0.2513	0.3800	0.2513	0.2513
u_7	0.4200	0.4103	0.4093	0.3200	0.2513	0.2513	0.3600	0.2513	0.2513
u_8	0.4400	0.4223	0.4520	0.3400	0.2513	0.2513	0.3400	0.2513	0.2513
u_9	0.4600	0.4228	0.4540	0.3600	0.2513	0.2932	0.3200	0.2513	0.2513
u_{10}	0.4800	0.4231	0.3777	0.3800	0.2513	0.2650	0.3000	0.2513	0.2513
Wt (lb)	2.1840	1.9174	1.9409	1.9600	1.5466	1.5629	2.1840	1.4991	1.5046
Violation					0.03			0.58	
Time (sec)		25.45	67.87		32.81	14.42		28.27	25.80
Time		1.16	2.42		1.17	2.06		1.28	1.98

[a]Results obtained with finite-dimensional method [Feng and co-workers, 1]

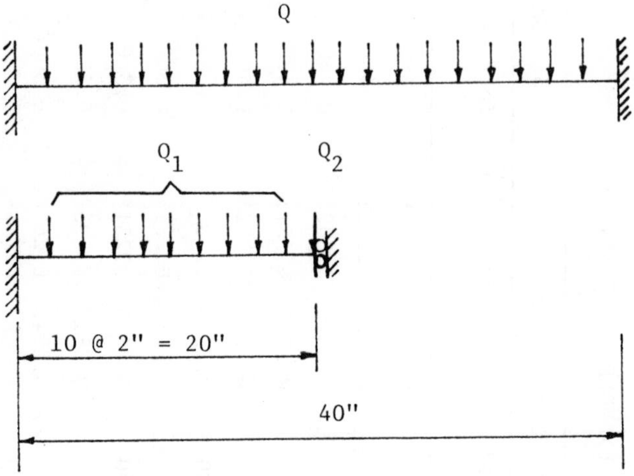

$\alpha = 0.3$, $E = 30 \times 10^6$ psi, $\gamma = 0.28$ lb/in^3

$\dot{z}(x,0) = z(x,0) = 0$

$$Q = \begin{cases} 9 \sin (80\,\pi t) \text{ lb/in}, & 0 \le t \le 0.0125 \text{ sec} \\ 0 \text{ lb/in}, & 0.0125 \le t \le 0.0150 \text{ sec} \end{cases}$$

$$Q_1 = \begin{cases} 18 \sin (80\,\pi t) \text{ lb}, & 0 \le t \le 0.0125 \text{ sec} \\ 0 \text{ lb}, & 0.0125 \le t \le 0.0150 \text{ sec} \end{cases}$$

$$Q_2 = \begin{cases} 9 \sin (80\,\pi t) \text{ lb}, & 0 \le t \le 0.0125 \text{ sec} \\ 0 \text{ lb}, & 0.0125 \le t \le 0.0150 \text{ sec} \end{cases}$$

Fig. 5. Fixed-fixed beam subjected to uniformly distributed load with $\alpha = 0.3$, $E = 30 \times 10^6$ psi, $\gamma = 0.28$ lb/in.^{3}, $\dot{z}(x, 0) = z(x, 0) = 0$, and Q, Q_1, and Q_2 as listed in Fig. 3.

The remaining problem formulation and equations required for implementation of the steepest descent method stay the same. Implementation thus required only minor modifications in the computer program used to solve the simply supported problem.

A clamped beam is loaded as shown in Fig.5. Design constraint parameters are $d = 0.1$ in and $u_0 = 0.2513$ in.^{2}. Only half the beam is modeled, as in Example 1, but with one end of the half beam fixed. A 19-degree-of-freedom finite-element model was employed and five eigenvectors were used for modal analysis. The first design resulted in $\|\delta u^1\| = 1.252$ without violating any constraints. The displacement constraint was first violated in the seventh iteration and remained active during the design process, except in iterations 12, 16, 25, and 26. The convergence measure was $\|\delta u^1\| = 0.0235$ in the final design shown in Table 1. The design weight history is shown in Fig. 8 in the next example.

C. Example 3: A Cantilevered Beam

For completeness, a cantilevered beam of Fig. 6 is treated.

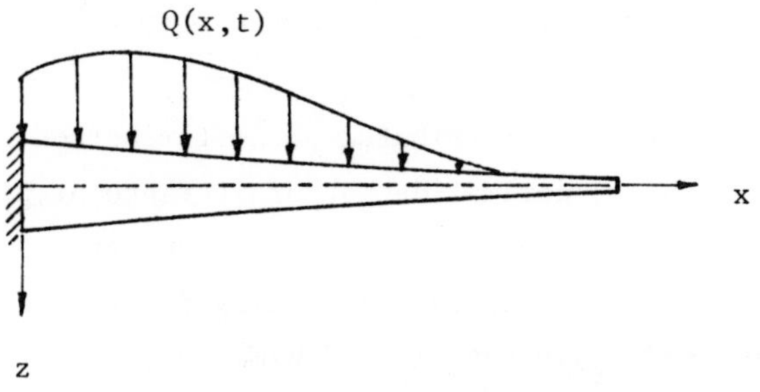

Fig. 6. Cantilevered Beam with Dynamic Load

Here, the boundary conditions of Eq. (49) are changed to

$$\left.\begin{aligned} z(0, t) &= \frac{\partial z}{\partial x}(0, t) = 0 \\ \frac{\partial^2 z}{\partial x^2}(a, t) &= \frac{\partial^3 z}{\partial x^3}(a, t) = 0 \end{aligned}\right\} \quad \text{in } T. \tag{60}$$

All other equations in the formulation and solution remain the same. Only minor changes in the computer program are required to solve this problem.

The length of a cantilevered beam is chosen to be half the length of the beams previously discussed. The cantilevered beam is partitioned into 10 elements for analysis, so it has 20 degrees of freedom. It is subjected to the distributed load shown in Fig. 7. Constraint parameters are as in Example 1. Five eigenvectors were used in modal analysis. At the initial design, $\|\delta u^1\| = 1.208$. At the final iteration, $\|\delta u^1\| = 0.147$. About half the beam, at the free end, reached the lower design variable bound before the displacement constraint was violated in the eighth design. This constraint violation was corrected to 0.58%. Design results are listed in Table 1 and the weight history is shown in Fig. 8.

V. DYNAMIC OPTIMIZATION OF RECTANGULAR PLATES

To illustrate the distributed parameter method on a higher-dimensional problem, optimal design of rectangular plates (two space dimensions) is considered. Simply supported and clamped plates are subjected to uniform spatially distributed, time-varying loads and their transient dynamic response is constrained. One seeks to determine the thickness variation $u(x_1, x_2)$ over the plate to minimize weight, subject to dynamic response

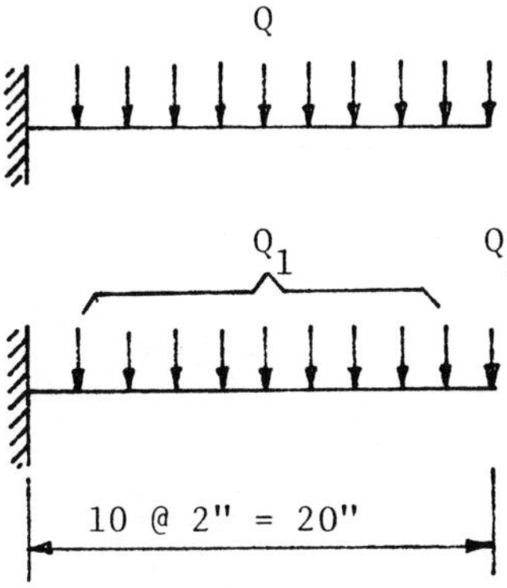

$\alpha = 0.3$, $E = 30 \times 10^6$ psi, $\gamma = 0.28$ lb/in^3

$\dot{z}(x,0) = z(x,0) = 0$

$$Q = \begin{cases} 9 \sin (80\,\pi t) \text{ lb/in,} & 0 \le t \le 0.0125 \text{ sec} \\ 0 \text{ lb/in,} & 0.0125 \le t \le 0.0150 \text{ sec} \end{cases}$$

$$Q_1 = \begin{cases} 18 \sin (80\,\pi t) \text{ lb,} & 0 \le t \le 0.0125 \text{ sec} \\ 0 \text{ lb,} & 0.0125 \le t \le 0.0150 \text{ sec} \end{cases}$$

$$Q_2 = \begin{cases} 9 \sin (80\,\pi t) \text{ lb,} & 0 \le t \le 0.0125 \text{ sec} \\ 0 \text{ lb,} & 0.0125 \le t \le 0.0150 \text{ sec} \end{cases}$$

Fig. 7. Cantilevered beam subjected to uniformly distributed load with $\alpha = 0.3$, $E = 30 \times 10^6$ psi, $\gamma = 0.28$ lb/in.^{3}, $\dot{z}(x, 0) = z(x, 0) = 0$, and values of Q, Q_1 and Q_2 as shown in Fig. 3.

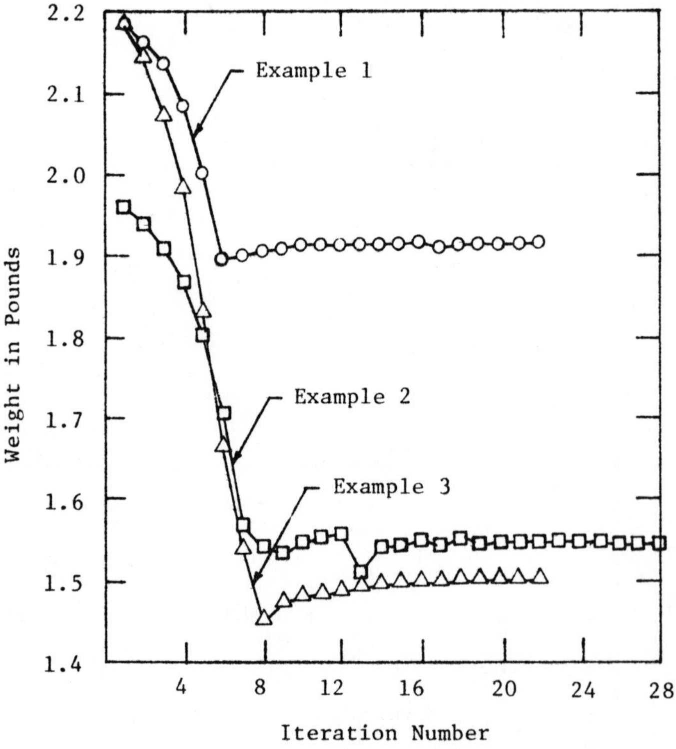

Fig. 8. Weight-Iteration Curves of Examples 1-3.

constraints.

A. Example 4: A Simply Supported Plate

For the simply supported rectangualr plate of Fig. 9, one wishes to choose the thickness variation $u(x_1, x_2)$ to minimize weight, subject to the conditions that displacement is within prescribed bounds and the thickness is bounded uniformly away from zero. Using subscript notation for partial differentiation, the system dynamic equations in this case may be written as [9]

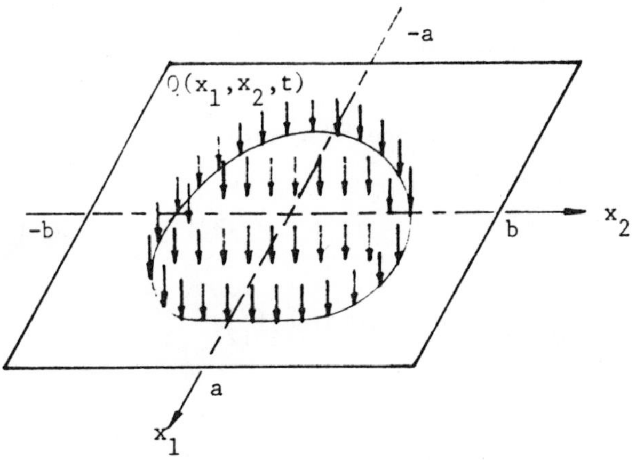

Fig. 9. A Rectangular Plate with Dynamic Load.

$$[D(u)(z_{11} + \nu z_{22})]_{11} + [D(u)(z_{22} + \nu z_{11})]_{22}$$

$$+ [2(1-\nu)D(u)z_{12}]_{12} + \rho u\ddot{z} = Q(x_1, x_2, t) \quad \text{in} \quad \Omega \times T, \tag{61a}$$

with boundary conditions

$$z(x_1, x_2, t) = 0 \tag{61b}$$

$$z_{11}(\pm a, x_2, t) = 0 \qquad \text{in} \quad \Gamma \times T \tag{61c}$$

$$z_{22}(x_1, \pm b, t) = 0 \tag{61d}$$

and initial conditions

$$z(x_1, x_2, 0) = 0 \tag{61e}$$

$$\text{in} \quad \Omega.$$

$$\dot{z}(x_1, x_2, 0) = 0 \tag{61f}$$

Here, $D(u) = \dfrac{Eu^3}{12(1 - \nu)^2}$.

One may verify that the operators

$$Lz \equiv [D(u)(z_{11} + \nu z_{22})]_{11} + [D(u)(z_{22} + \nu z_{11})]_{22}$$

$$+ 2[(1 - \nu)D(u)z_{12}]_{12}$$

and

$$Mz \equiv \rho u\ddot{z}$$

are symmetric, in the sense defined in Section II. Further,

$$\Delta_u(Lz) = [D'(u)(z_{11} + \nu z_{22})]_{11} + [D'(u)(z_{22} + \nu z_{11})]_{22}$$

$$+ 2[(1 - \nu)D'(u)z_{12}]_{12}$$

and

$$\Delta_u(Mz) = \rho\delta u\ddot{z}.$$

Integration by parts in Eq. (21) yields

$$\iint_\Omega \lambda\Delta_u\, Lz\, dx = \iint_\Omega [D'(u)[\lambda_{11}z_{11} + \nu\lambda_{11}z_{22} + \lambda_{22}z_{22}$$

$$+ \nu\lambda_{22}z_{11}] + 2(1 - \nu)D'(u)\lambda_{12}z_{12}]\ \delta u\, dx$$

$$\equiv \iint_\Omega \hat{L}(x, \lambda)\ \delta u\, dx \tag{62}$$

and

$$\iint_\Omega \int_0^\tau \lambda\ \Delta u(Mz)\ dx\, dt = \iint_\Omega \int_0^\tau [-\rho\dot{\lambda}\dot{z}]\ \delta u\, dx\, dt$$

$$\equiv \iint_\Omega \int_0^\tau M(z, \lambda)\ \delta u\, dx\, dt. \tag{63}$$

The weight to be minimized is

$$\psi_0 = \iint_\Omega \lambda u\, dx \qquad \text{or} \qquad \psi_0 = \int_0^\tau \iint_\Omega \frac{\gamma}{\tau} u\, dx\, dt, \tag{64}$$

where γ is the material weight density. The displacement constraint

$$|z(x_1, x_2, t)| \leq d \qquad \text{in} \quad \Omega \times T$$

is transformed to functional form as

$$\psi_1 = \int_0^\tau \iint_\Omega [(|z| - d) + \Big||z| - d\Big|] \, dx \, dt = 0. \tag{65}$$

Finally, a lower bound is placed on the plate thickness

$$\phi_1 = -u(x_1, x_2) + u_0 \leq 0 \qquad \text{in} \quad \Omega. \tag{66}$$

The functional constraint of Eq. (65) requires solution of an adjoint problem; namely, Eq. (19). By inspection, this is just Eq. (61), with the applied load replaced by

$$\overline{Q} = \frac{\partial f_1}{\partial z} = \operatorname{sgn}(z)[1 + \operatorname{sgn}(|z| - d)] \qquad \text{in} \quad \Omega. \tag{67}$$

and with initial conditions replaced by terminal conditions

$$\lambda(x_1, x_2, \tau) = \dot{\lambda}(x_1, x_2, \tau) = 0 \qquad \text{in} \quad \Omega. \tag{68}$$

By inspection $\Lambda^{\psi_0} = \gamma$ and $\dfrac{\partial \phi_1}{\partial u} = -1$. From Eqs. (24), (62), (63), and (65):

$$\Lambda^{\psi_1}(x_1, x_2) = -\int_0^\tau [D'(u)[\lambda_{11}z_{11} + \nu\lambda_{11}z_{22} + \lambda_{22}z_{22} + \nu\lambda_{22}z_{11}]$$

$$+ 2(1 - \nu)D'(u)\lambda_{12}z_{12} - \rho\dot{\lambda}\dot{z}] \, dt. \tag{69}$$

All information is now available to implement the optimization algorithm of Section III.

As a numerical example, a square plate $a = b = 20$ in. is

chosen, with material properties $E = 30 \times 10^6$ psi, $\nu = 0.3$, and $\gamma = 0.28$ lb/in.^{3}. The plate is loaded with a time-varying load that is uniformly distributed over Ω:

$$Q(x_1, x_2, t) = \begin{cases} \sin(80\,\pi t) \text{ lb/in.}^2, & 0 \leq t \leq 0.0125 \text{ sec} \\ 0 \text{ lb/in.}^2, & 0.0125 \leq t \leq 0.0175. \end{cases}$$

Constraint parameters are $d = 0.4$ in and $u_0 = 0.0625$ in.

In Fig. 10, one quarter of the plate is shown divided into 25 finite elements for analysis. The finite-element model had 75 degrees of freedom, and six eigenfunctions were employed for modal analysis. The load was discretized in the same manner as in the beam problems. The initial design estimate weighed 20.61 lb. The maximum deflection was 0.428 in. at $t = 0.0165$ sec. In the 12th design iteration, the constraint was completely satisfied. Numerical results are shown in Table 2. Note that the thickness of the plate is symmetric with respect to the bisector of the x-y axes in Fig. 10. A sketch of the resultant thickness contours is shown in Fig. 11.

While convergence properties for this problem were not as good as with the beam problems of Section IV, convergence from several design estimates led to the same design weight given in Table 2.

B. Example 5: A Clamped Plate

The plate optimization problem for a clamped plate follows directly from the preceding analysis. The only modification required is to change the boundary conditions in Eq. (61) to

$$z(x_1, x_2, t) = 0 \tag{70a}$$

TABLE 2

INITIAL AND FINAL DESIGNS OF EXAMPLES 4 AND 5

Examples:	4. Simply Supported Plate		5. Clamped Plate	
Design (in^2)	Initial	Final	Initial	Final
u_1	0.3600	0.3993	0.1100	0.0625
u_2	0.2600	0.2791	0.1100	0.0625
u_3	0.1600	0.1316	0.2200	0.1475
u_4	0.1100	0.0625	0.4200	0.3482
u_5	0.1100	0.0911	0.6200	0.5554
u_6	0.2600	0.2635	0.1100	0.0625
u_7	0.1600	0.1020	0.1100	0.0628
u_8	0.1100	0.0788	0.1200	0.0779
u_9	0.1100	0.1335	0.2200	0.2641
u_{10}	0.1600	0.1330	0.1100	0.0685
u_{11}	0.1600	0.1856	0.1100	0.0797
u_{12}	0.1600	0.1892	0.1200	0.1438
u_{13}	0.2600	0.2865	0.2200	0.1666
u_{14}	0.2600	0.2816	0.3200	0.2739
u_{15}	0.3600	0.3708	0.4200	0.3665
Wt (lb)	0.2061	20.12	25.58	21.31
Violation %	7.03	0	0	1.51
Time (sec)		121.81		133.57
Time		10.15		9.54

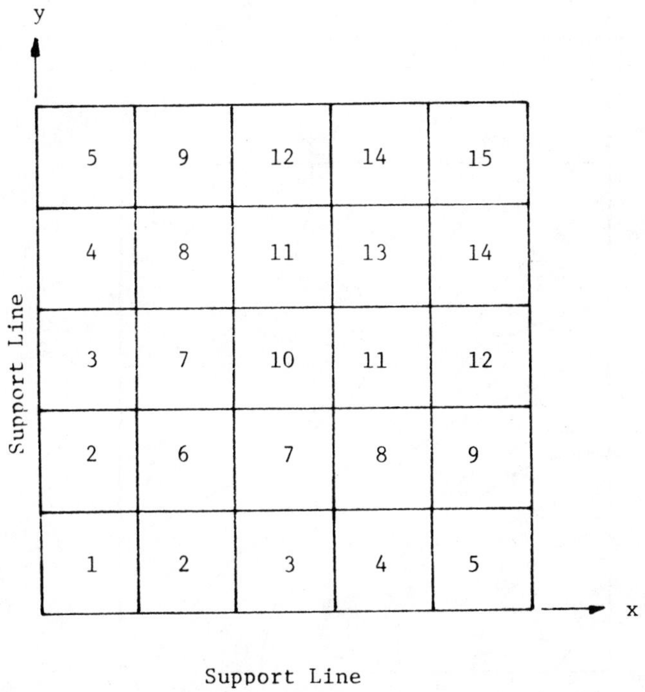

Fig. 10. Finite-Element Model of One-Quarter of a Square Plate.

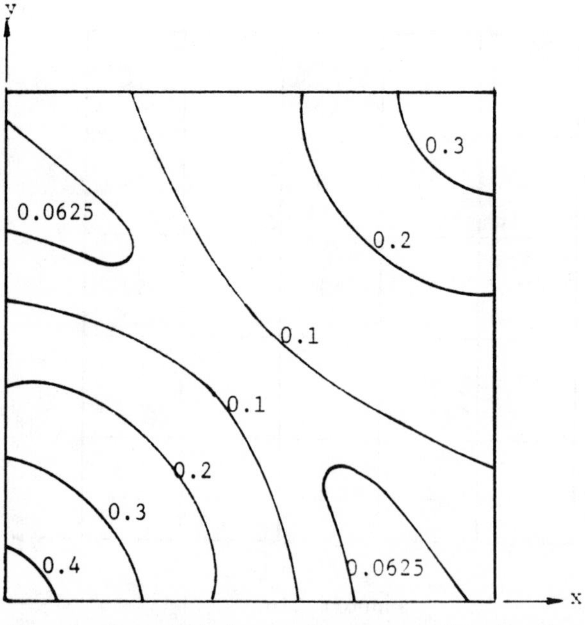

Fig. 11. Optimum Thickness Contours of a Simply Supported Plate.

and

$$\frac{\partial z}{\partial x_1}(\pm a,\ x_2,\ t) = 0 \qquad \text{(70b)}$$

$$\text{in} \quad \Gamma \times T.$$

$$\frac{\partial z}{\partial x_2}(x_1,\ \pm b,\ t) = 0 \qquad \text{(70c)}$$

The remaining problem formulation and equations required for implementation of the steepest descent method are the same as in Example 4.

As a numerical example, a plate with the same planar dimensions and material properties is subjected to the same load as the simply supported plate in Example 4. Constraint parameters are $d = 0.1$ in. and $u_0 = 0.0624$ in. In Fig. 10, one quarter of the plate is shown divided into 25 elements for analysis. The resulting finite-element model had 65 degrees of freedom and six eigenfunctions were used for analysis. The initial design estimate had a weight of 25.58 lb. Constraint violation was found to be about 4.6% in the second iteration. The optimum design was found to have a design weight of 21.30 lb, with a maximum displacement constraint violation of 1.5%. Numerical results are shown in Table 2 and the resultant thickness contours are shown in Fig. 12. Numerical calculations were more stable in this problem than in Example 4. Several initial design estimates led to the same optimum.

VI. SUMMARY AND CONCLUSIONS

The distributed parameter optimal design problem formulated herein takes advantage of the fact that the design variable is independent of time. A sensitivity analysis method is developed, which provides cost and constraint funcitonal sensitivity in the

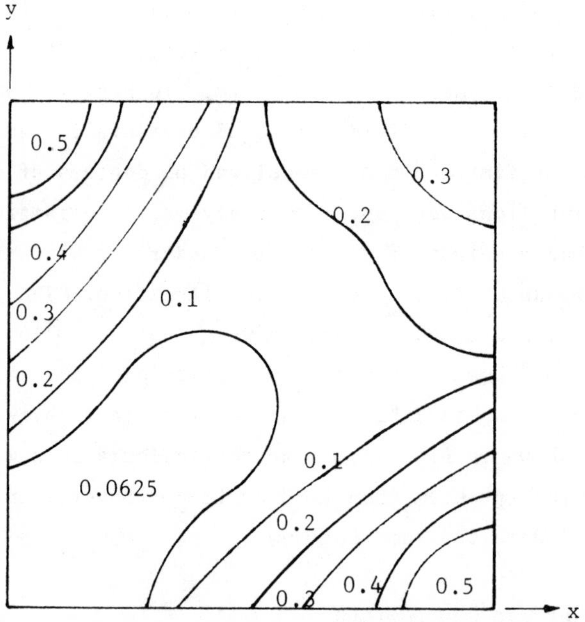

Fig. 12. Optimum Thickness Contours of a Clamped Plate.

time-independent design space. The resulting design sensitivity data allow for direct application of an effective generalized steepest descent computational algorithm.

The optimal design algorithm is applied to beams and plates, under dynamic load. These problems represent one and two space dimensions for design distribution, respectively, each with an additional time dimension. Both classes of problems are solved by the algorithm presented herein, with typical computing times of 30 sec for a beam problem and 130 sec for a plate problem; all calculations were performed as in IBM 360/65 computer.

A special feature of the method presented is the ability to treat state variable inequality constraints $\eta(z) \leq 0$ through the equivalent functional constraint

$$\psi = \int_0^\tau \int_\Omega (\eta + |\eta|)\, dx\, dt = 0.$$

This feature is inherent in both the beam and plate problems treated herein. This technique was successfully applied in both the beam and plate problems, but with some minor computational difficulty in the case of the plate. The computational difficulty arises due to the fact that if the constraint $\eta \leq 0$ is very nearly satisfied, then the function $\eta + |\eta|$ is nonzero over only a very small subset of $\Omega \times T$. Hence, the sensitivity calculations for $\delta\psi$ are subject to numerical error and the parameter $M_{\psi\psi}$ in the numerical algorithm is very small, leading to oscillation in the optimization algorithm. This difficulty only limited convergence of the algorithm to the precise optimum. It did not preclude convergence to a practical approximately optimum design.

REFERENCES

1. T.T. FENG, J.S. ARORA, and E.J. HAUG, Internat. J. Numer.

Methods Engrg., to appear, (1976).

2. J.L. LIONS, "Optimal Control of Systems Governed by Partial Differential Equations." Springer-Verlag, Berlin and New York, 1971.

3. A.G. BUTKOVSKIY, "Distributed Control Systems" Amer. Elsevier, New York, 1969.

4. IFAC Symp. Control of Distributed Parameter Systems, Preprints, University of Calgory, Banff, Canada, 1971.

5. E.J. HAUG, K.C. PAN, and T.D. STREETER, Internat. J. Numer. Methods Engrg. 9, (3), 469-667, (1975).

6. S.G. MIKHLIN, "Mathematical Physics, An Advanced Course", North-Holland Publ. Amsterdam, 1970.

7. A.E. BRYSON, and Y.C. HO, "Applied Optimal Control", Ginn, Boston, Massachusetts, 1969.

8. E.J. HAUG, "Computer Aided Design of Mechanical Systems, Engineering Design Handbook", AMCP 706-192. Army Material Command, Washington, D.C., 1973.

9. C.T. WANG, "Applied Elasticity", McGraw-Hill, New York, 1953.

10. J.S. PRZEMIENICKI, "Theory of Matrix Structural Analysis", McGraw-Hill, New York, 1968.

11. K.J. BATHE, and E.L. WILSON, J. Engrg. Mech. Div., ASCE, Vol. 98, (EM6), 1471-1485, 1972.

OPTIMAL OBSERVERS FOR CONTINUOUS TIME LINEAR STOCHASTIC SYSTEMS

JOHN F. YOCUM, JR.

Space and Communications Group
Hughes Aircraft Company
El Segundo, California

I. INTRODUCTION

The problem of state estimation in linear systems is of fundamental importance, not only in its own right, but also as a factor in feedback control loop design, particularly in problems with time-varying systems or with highly coupled multi-input/output systems. Kalman [1] in 1960 solved the optimal estimation problem for an n-dimension linear system with white stochastic inputs and noisy measurements. He found that an n-dimensional filter was required. On the other extreme, in 1964 an important paper by Luenberger [2] considered the state estimation problem for an n-dimensional noise-free linear system with m noise-free outputs. The principal result in that paper is that a non-unique $(n - m)$-dimensional filter (called an "observer") could be constructed whose output could be linearly combined with the m measurements to provide a state estimate which converged asymptotically to the true state. Moreover, the $n - m$ eigenvalues of the filter could be arbitrarily chosen. These generalizations apply to discrete-time as well as to continuous-time systems.

It is natural to expect that if noise were introduced into the heretofore deterministic observer problem of Luenberger, one of the nonunique Luenberger observers would be better than the others in the sense that its mean square estimation error would be less. That is to say, of the infinite number of $(n - m)$-dimensional deterministic observers for a given system, one, or possibly an identifiable class, must excel the others in terms of its mean square estimation error. This one observer, or class of observers, could be termed the optimal $n - m$ order observer(s) for that system. In fact, Luenberger showed that nonunique observers of any order between $n - m$ and n can be constructed. Consequently, one may postulate the existence of several optimal observers, ranging in dimension from $n - m$ to n. The n-dimension observer would be a Kalman filter.[†] Novak and Leondes in [3], (see also Novak, Yocum and Leondes [4, 5, 6, 7]) show how to construct optimal observers for discrete-time systems. This chapter considers continuous-time systems and derives the optimal observer of any given dimension in the range $n - m$ to n. The form of the solution is very similar to a Kalman filter and, in fact, is identical to the Kalman algorithm for the n-dimensional observer.

II. PROBLEM FORMULATION

In this section we define the system under consideration and give a precise statement of the estimation problem to be solved. Attention is focused on the class of continuous-time linear dynamical systems and measurements described by:[‡]

[†] If some of the measurements are noise free, then the dimension of state equation, and hence the Kalman filter dimension, could be reduced by eliminating some of the state variables in favor of the noise-free measurements which are known. Tse and Athans [8,9] made this observation less directly. We assume here that such reduction has already been done and all remaining measurements are noisy.

[‡] In this chapter, all differential equations of the form of Eq.(1), where $w(t)$ is a vector white-noise process, are to be taken as shorthand expressions for stochastic differential equations of the form

$$dx(t) = F(t)x(t)\,dt + G(t)u(t)\,dt + M(t)v(t)\,dt + dw(t),$$

where $dw(t)$ is an appropriate vector Brownian motion process, and are to be interpreted in the sense of Ito.

System State Equation:

$$\dot{x}(t) = F(t)x(t) + G(t)u(t) + M(t)v(t) + w(t); \tag{1}$$

Measurement Equation:

$$\begin{Bmatrix} y_1(t) \\ \\ y_2(t) \end{Bmatrix} = \begin{bmatrix} I_{m_1} & 0 \\ & \\ H_{21}(t) & H_{22}(t) \end{bmatrix} \begin{Bmatrix} x_1(t) \\ \\ x_2(t) \end{Bmatrix} + \begin{bmatrix} L_1(t) \\ \\ L_2(t) \end{bmatrix} v(t) + \begin{Bmatrix} 0 \\ \\ e_2(t) \end{Bmatrix}, \tag{2}$$

where the following definitions apply: $(\dot{\ })$ implies $d/dt(\ )$; $x(t) = n \times 1$ state vector; $E\{x(0)\} = \bar{x}_0$ and $E\{x(0)x(0)^T\} = P_{xx}(0)$, where $\bar{x}_0$, $P_{xx}(0)$ are given and the separation of x into x_1 and x_2 is based on the measurements as explained in comment (ii) below; $u(t) = r \times 1$ control vector;

$$\begin{Bmatrix} y_1(t) \\ \\ y_2(t) \end{Bmatrix} = \begin{Bmatrix} m_1 \times 1 \text{ measurement vector} \\ \\ m_2 \times 1 \text{ measurement vector} \end{Bmatrix} \text{where } (m_1 + m_2 \triangleq m).$$

In addition $w(t) = n \times 1$ zero mean white-noise process, independent of $\bar{x}_0$ and with autocorrelation $E\{w(t)w(\tau)^T\} = W(t)\delta(t-\tau)$; $e_2(t) = m_2 \times 1$ zero mean white-noise process, independent of $\bar{x}_0$ and $w(t)$ and with autocorrelation $E\{e_2(t)e_2(\tau)^T\} = E(t)\delta(t-\tau)$. Finally $v(t) = q \times 1$ Markov one (colored noise) process which satisfies

$$\dot{v}(t) = \Phi(t)v(t) + n(t); \qquad E\{v(0)\} = 0; \tag{3}$$

in which

$$E\{v(0)[v^T(0), \quad x^T(0)]\} = [P_{vv}(0), \quad P_{vx}(0)]$$

with $P_{vv}(0)$ and $P_{vx}(0)$ given; and $n(t) = q\times 1$ zero mean white-noise process, independent of $\bar{x}_0$, $w(t)$, and $e_2(t)$, and with autocorrelation $E\{n(t)n(\tau)^T\} = N(t)\delta(t-\tau)$. The matrices $F(t)$, $G(t)$, $M(t)$, $H_{21}(t)$, $H_{22}(t)$, $L_1(t)$, $L_2(t)$, $W(t)$, $E(t)$, and $N(t)$ are known matrices of appropriate dimensions and $I_{m_1} = m_1 \times m_1$ unit matrix.

The estimation problem is to construct an optimal linear estimator of the form

State: $$\dot{\xi}(t) = A(t)\xi(t) + [K'_{21}(t), K_{22}(t)]\begin{Bmatrix} y_1(t) \\ y_2(t) \end{Bmatrix} + B(t)u(t), \tag{4}$$

Estimator: $$\hat{x}(t) = C(t)\xi(t) + D(t)y_1(t) \tag{5}$$

where ξ is of dimension p (p is specified beforehand and may lie in the range $n - m_1$ to $n - m_1 + q$). The estimator is optimal in the sense that the matrices A, K'_{21}, K_{22}, B, C, and D are chosen to minimize the following quadratic cost function

$$J = E\{[x(t) - \hat{x}(t)]^T Q'(t)[x(t) - \hat{x}(t)]\}. \tag{6}$$

This problem differs somewhat from the usual linear estimation problem. The following comments may help to explain these differences.

(i) The measurement vector is separated into two portions. the measurements $y_2(t)$ may be white-noise processes with infinite variance due to additive white-noise $e_2(t)$. Direct feedthrough of $y_2(t)$ into the estimate would result in $J \to \infty$, and is therefore not permitted. By contrast, $y_1(t)$ is a Markov one process with finite variance and its direct feedthrough into $\hat{x}$ allows for a bounded cost and, hence, optimization. In a given application it may happen that $e_2(t) \equiv 0$ and $y_2(t)$ is also a Markov one process. In this case, the distinction between $y_1(t)$ and $y_2(t)$ is entirely up to the designer. As is apparent in Eq. (5), however, $y_1(t)$ enters the estimate additively and as a rule should contain the most accurate measurements.

(ii) The measurements $y_1(t)$ are assumed to be direct measurements of the first m_1 states x_1 rather than a linear combination of all states. As will be seen, this assumption simplifies the resulting derivation considerably. The assumption does not compromise generality since any system of the form $y_1 = H_1^* x^*$, where H_1^* is of full rank, can be cast into the desired form by a change of variable

$$\begin{Bmatrix} x_1 \\ x_2 \end{Bmatrix} = \begin{bmatrix} H_1^* \\ T \end{bmatrix} x^*$$

in which T is any $(n - m_1) \times n$ matrix such that

$$\begin{bmatrix} H_1^* \\ T \end{bmatrix}$$

is nonsingular.

(iii) The system state and measurement equations allow for colored noise inputs via $v(t)$, as well as the usual white-noise inputs. In most applications disturbances are best described as colored, rather than white, noises. Although initial inspection of Eqs. (1) and (2) seems to indicate a restriction that the same colored noise drives both the plant and the measurements, this is not necessarily the case. By appropriate choice of $\phi(t)$ and $n(t)$, various "state" elements within $v(t)$ can be made independent of one another. Then proper choice of $M(t)$ and $L(t)$ will result in completely independent noises in the two equations.

(iv) The form of the estimator equations (4) and (5) is assumed at the outset to be a linear system, because linear systems are easiest to realize. Kalman has shown that under rather general conditions, the optimal solution with $p = n$ is linear, even when the form is not presupposed.

(v) The estimator equation (5) allows for direct feedthrough of the measurement $y_1(t)$. No direct feedthrough appears in any of Kalman's work. The feedthrough concept is motivated by Luenberger's [2, 10] work on observers.

III. STRUCTURE OF THE OBSERVER--DETERMINISTIC CONSIDERATIONS

In this section our goal is to unmask the basic structure of an observer, using primarily deterministic considerations. With the exception of a few initial observations, the approach followed is motivated by the work of Luenberger [2]. The results are applicable to all observers, both optimal and nonoptimal. Choice of the optimal observer is deferred until Section IV.

To begin we reformulate the state equations by eliminating $x_1(t)$ in favor of $y_1(t)$ and $v(t)$. From Eq. (2),

$$x_1(t) = y_1(t) - L_1(t)v(t) \tag{7}$$

so that

$$\dot{x}_1(t) = \dot{y}_1(t) - \dot{L}_1(t)v(t) - L_1(t)\dot{v}(t). \tag{8}$$

Substituting these into the state and measurement equations (1) and (2), and taking account of Eq. (3), we can obtain

$$\left\{\begin{array}{c} \dot{v} \\ \hline \dot{x}_2 \end{array}\right\} = \left[\begin{array}{c|c} \Phi & 0 \\ \hline M_2 - F_{21}L_1 & F_{22} \end{array}\right]\left\{\begin{array}{c} v \\ \hline x_2 \end{array}\right\} + \left\{\begin{array}{c} 0 \\ \hline G_2u + F_{21}y_1 \end{array}\right\} + \left\{\begin{array}{c} n \\ \hline w_2 \end{array}\right\} \tag{9}$$

and

$$\left\{\begin{array}{c} \dot{y}_1 - F_{11}y_1 - G_1u \\ \hline y_2 - H_{21}y_1 \end{array}\right\}$$

$$= \left[\begin{array}{c|c} \dot{L}_1 - F_{11}L_1 + L_1\Phi + M_1 & F_{12} \\ \hline L_2 - H_{21}L_1 & H_{22} \end{array}\right]\left\{\begin{array}{c} v \\ \hline x_2 \end{array}\right\} + \left\{\begin{array}{c} L_1n + w_1 \\ \hline e_2 \end{array}\right\} \tag{10}$$

in which the dependence on time has been suppressed to simplify notation and the following matrix partitioning has been adopted:

$$M = \left[\begin{array}{c} M_1 \\ \hline M_2 \end{array}\right]\begin{array}{l} \updownarrow m_1 \\ \updownarrow n-m_1 \end{array}, \quad |\leftarrow q \rightarrow| \qquad F = \left[\begin{array}{c|c} F_{11} & F_{12} \\ \hline F_{21} & F_{22} \end{array}\right]\begin{array}{l} \updownarrow m_1 \\ \updownarrow n-m_1 \end{array}, \quad |\leftarrow m_1 \rightarrow|\leftarrow n-m_1 \rightarrow|$$

$$G = \left[\begin{array}{c} G_1 \\ \hline G_2 \end{array}\right]\begin{array}{l} \updownarrow m_1 \\ \updownarrow n-m_1 \end{array}, \qquad w = \left\{\begin{array}{c} w_1 \\ \hline w_2 \end{array}\right\}\begin{array}{l} \updownarrow m_1 \\ \updownarrow n-m_1 \end{array}.$$

Equation (9) may be viewed as a new state equation with the $(q + n - m_1)$-dimensional state vector

$$\begin{Bmatrix} v \\ \hline x_2 \end{Bmatrix} \begin{matrix} \uparrow \\ q \\ \downarrow \\ \uparrow \\ n-m_1 \\ \downarrow \end{matrix} \tag{11}$$

while Eq. (10) is viewed as a new measurement equation with derived measurement vector

$$z = \begin{Bmatrix} z_1 \\ \hline z_2 \end{Bmatrix} = \begin{Bmatrix} \dot{y}_1 - F_{11}y_1 - G_1u \\ \hline y_2 - H_{21}y_1 \end{Bmatrix}. \tag{12}$$

By making the following notational definitions:

$$H = [H_1 \quad H_2] = \left[\begin{array}{c|c} \dot{L}_1 - F_{11}L_1 + L_1\Phi + M_1 & F_{12} \\ L_2 - H_{21}L_1 & H_{22} \end{array}\right] \tag{13a}$$

and

$$\nu = \begin{Bmatrix} L_1 n - w_1 \\ e_2 \end{Bmatrix} \tag{13b}$$

we can write Eq. (10) simply as

$$z = H \begin{Bmatrix} v \\ x_2 \end{Bmatrix} + \nu = H_1 v + H_2 x_2 + \nu. \tag{10a}$$

In this form the system is recognized as being linear with additive white-noise plus known inputs, and with linear measurements corrupted by additive white noise. This is a form to which the Kalman filter is directly applicable. The corresponding Kalman filter is of dimension $q + n - m_1$, and estimates the colored noise state $v(t)$ as well as the remaining state vector $x_2(t)$.

As an alternative appraoch, we note from Eq. (3) that $v(t)$ is a zero mean process. Hence, rather than devote q states of a filter to estimating $v(t)$, we take as our estimate

$$\hat{v}(t) = E\{v(t)\} \equiv 0, \qquad t \geq 0. \tag{14}$$

As a direct consequence of this assumption and Eq. (7), we have

$$\hat{x}_1 = y_1 - L_1\hat{v} = y_1, \tag{15}$$

and it remains only to construct a filter of dimension $n - m_1$ to estimate the $n - m_1$ states of x_2. For this task we assume a general linear estimator of the form

$$\begin{aligned}\dot{\hat{x}}_2(t) = A(t)\hat{x}_2(t) &+ K_2(t)z(t) + B'(t)u(t) \\ &+ D'(t)y_1(t)\end{aligned} \tag{16}$$

and, following Luenberger, form the differential equation for the estimation error $\Delta x_2 \triangleq x_2 - \hat{x}_2$, by substracting Eq. (16) from the lower portion of Eq. (9). Finally, Eq. (10a) is used to eliminate z and the result is

$$\begin{aligned}\Delta\dot{x}_2 = {} & A\Delta x_2 + w_2 - K_2\nu + (M_2 - F_{21}L_1 - K_2H_1)v \\ & + (F_{22} - A - K_2H_2)x_2 + (G_2 - B')u + (F_{21} - D')y_1.\end{aligned} \tag{17}$$

We require the estimate to converge to $x_2(t)$ independently of the time histories of $y_1(t)$, $u(t)$, or $x_2(t)$. Necessary conditions for such convergence are that the coefficients of u, y_1, and x_2 be independently zero. Thus

$$B' = G_2, \tag{18}$$

$$D' = F_{21}, \tag{19}$$

$$A = F_{22} - K_2H_2. \tag{20}$$

The estimation error cannot be required to be independent of the colored noise $v(t)$ so that its coefficient is not set to zero.

The resulting estimate and estimation error take the form

$$\text{Estimate:}\quad \dot{\hat{x}}_2 = F_{22}\hat{x}_2 + K_2(z-H_2\hat{x}_2) + G_2u + F_{21}y_1, \tag{21}$$

$$\text{Estimation Error:}\quad \Delta\dot{x}_2 = (F_{22} - K_2H_2)\,\Delta x_2 + (M_2 - F_{21}L_1 - K_2H_1)v + w_2 - K_2\nu. \tag{22}$$

It is observed at this point that the estimator is described in terms of only a single unknown matrix K_2. As evidenced by Eq. (22), in the absence of noises v, w_2, and ν, any choice of K_2 which yields a stable matrix for $(F_{22} - K_2H_1)$ will result in an estimator whose error approaches zero. In Section IV we use stochastic considerations to select an optimal K_2 matrix.

IV. CHOICE OF OPTIMAL K_2 MATRIX

In this section we obtain an expression for the K_2 matrix which is optimal in the sense that it minimizes the cost

$$J = E\{\Delta x(t)^{\mathrm{T}}Q'(t)\;\Delta x(t)\}. \tag{6a}$$

First we rewrite this cost function by noting that from Eqs. (7) and (15)

$$\Delta x_1 \triangleq x_1 - \hat{x}_1 = -L_1 v$$

$$= -L_1(v - \hat{v}) = -L_1 \, \Delta v \tag{23}$$

since $\hat{v} \equiv 0$. Then, with $Q(t)$ defined as

$$Q(t) \triangleq \begin{bmatrix} -L_1(t) & 0 \\ 0 & I \end{bmatrix}^{\mathrm{T}} Q'(t) \begin{bmatrix} -L_1(t) & 0 \\ 0 & I \end{bmatrix} \tag{24}$$

the cost can be written as

$$J = E \begin{Bmatrix} \Delta v \\ \Delta x_2 \end{Bmatrix}^{\mathrm{T}} Q(t) \begin{Bmatrix} \Delta v \\ \Delta x_2 \end{Bmatrix}$$

$$= \mathrm{tr}\, E\, Q(t) \begin{Bmatrix} \Delta v \\ \Delta x_2 \end{Bmatrix} \begin{Bmatrix} \Delta v \\ \Delta x_2 \end{Bmatrix}^{\mathrm{T}}$$

$$= \mathrm{tr}\{Q(t)P(t)\} \tag{25}$$

in which

$$P(t) = E \begin{Bmatrix} \Delta v \\ \Delta x_2 \end{Bmatrix} \{\Delta v^{\mathrm{T}} \quad \Delta x_2^{\mathrm{T}}\} = \text{error covariance matrix} \tag{26}$$

and $\mathrm{tr}(\cdot)$ denotes the trace operator.

The optimization can be formulated as an optimal control probelm as follows. First we write the differential equation satisfied by $P(t)$. From Eqs. (3) and (21) we note that the vector $\begin{matrix}\Delta v\\ \Delta x_2\end{matrix}$ satisfies a linear differential equation driven by white noise

$$\left\{\begin{array}{c}\Delta \dot{v}\\ \hline \Delta \dot{x}_2\end{array}\right\} = \left[\begin{array}{c|c}\Phi & 0\\ \hline M_2 - F_{21}L_1 - K_2H_2 & F_{22} - K_2H_2\end{array}\right]\left\{\begin{array}{c}\Delta v\\ \hline \Delta x_2\end{array}\right\} + \left\{\begin{array}{c}n\\ \hline w_2 - K_2\nu\end{array}\right\}. \tag{27}$$

As is well known (see, for example, Meditch [11, pp. 143-148]) the variance for such an equation satisfies a matrix differential equation of the form

$$\dot{P}(t) = \psi(K_2, t)P(t) + P(t)\psi^T(K_2, t) + \Gamma(K_2, t), \tag{28}$$

where

$$\psi(K_2, t) = \left[\begin{array}{c|c}\Phi & 0\\ \hline M_2 - F_{21}L_1 - K_2H_2 & F_{22} - K_2H_2\end{array}\right] \tag{29}$$

and $\Gamma(K_2, t)$ is the covariance kernel of the white noise defined as

$$E\left(\left\{\begin{array}{c}n(t)\\ w_2(t) - K_2(t)\nu(t)\end{array}\right\}\left\{\begin{array}{c}n(\tau)\\ w_2(\tau) - K_2(\tau)\nu(\tau)\end{array}\right\}^T\right)$$

$$= \Gamma(K_2, t)\delta(t-\tau) \tag{30}$$

From the definition of ν in Eq. (13) and the various stochastic properties of $e_2(t)$, $w(t)$, and $n(t)$ defined in Section II,

$\Gamma(K_2, t)$ is found to be

$$\Gamma(K_2, t) = \left[\begin{array}{c|c} N & -[NL_1^T \;\; 0]K_2^T \\ \hline -K_2\begin{bmatrix} L_1N \\ 0 \end{bmatrix} & K_2\left[\begin{array}{c|c} W_1 + L_1NL_1^T & 0 \\ \hline 0 & E_2 \end{array}\right]K_2^T - K_2\begin{bmatrix} W_{12} \\ \hline 0 \end{bmatrix} - \begin{bmatrix} W_{12} \\ \hline 0 \end{bmatrix}^T K_2^T + W_2 \end{array}\right]. \tag{31}$$

Choice of the minimizing K_2 matrix can now be stated as an optimal control problem.

OPTIMAL CONTROL PROBLEM. Select the "control" matrix $K_2(t)$ for $t_0 \leq t \leq t_f$ to minimize the terminal time cost

$$J(t_f) = \mathrm{tr}[Q(t_f)P(t_f)] \tag{32}$$

subject to the constraint

$$\dot{P}(t) = \psi(K_2, t)P(t) + P(t)\psi^T(K_2, t) + \Gamma(K_2, t) \tag{28}$$

with initial condition

$$P(t_0) = P_0 \quad \text{given.} \tag{33}$$

The "state" and "control" in this terminal time optimal control problem are the matrices $P(t)$ and $K_2(t)$, respectively. Problems of this sort can be solved using the maximum principle (see Athans [12]) or by calculus of variations as is done by Yocum [5]. The result is, not surprisingly, given in terms of a two-point boundary value problem

$$K_2(t) = \left[\Lambda_{22}^{-1}(t)\Lambda_{21}(t) \;\middle|\; I\right]\left[P(t)H(t) + S(t)\right]$$
$$\times \left[\begin{array}{c|c} W_1(t) + L_1(t)N(t)L_1(t)^{\mathrm{T}} & 0 \\ \hline 0 & E_2(t) \end{array}\right]^{-1}, \tag{34}$$

where the "state" $P(t)$ and the "adjoint matrix" $\Lambda(t)$ are given as solutions to

$$\dot{P} = \psi P + P\psi^{\mathrm{T}} + \Gamma, \qquad P(t_0) = P_0, \tag{28a}$$

$$\dot{\Lambda} = -\psi^{\mathrm{T}}\Lambda - \Lambda\psi, \qquad \Lambda(t_f) = Q(t_f). \tag{35}$$

In Eq. (34) the matrix $S(t)$ is defined as

$$S(t) = \left[\begin{array}{c|c} NL_1^{\mathrm{T}} & 0 \\ \hline W_{12}^{\mathrm{T}} & 0 \end{array}\right]. \tag{36}$$

Solution of the matrix two-point boundary value problem in Eqs. (28) and (35) yields the time history for $K_2(t)$ over the interval $t_0 \leq t \leq t_f$ which results in minimum cost at time t_f but may not produce the best estimate at intermediate times. In a filtering situation, we are interested in the case where $t_0 \equiv t \equiv t_f$. Equation (35) reduces to

$$\Lambda(t) \equiv Q(t) \tag{37}$$

and the optimal K_2 is

$$K_2 = [Q_{22}^{-1} Q_{21} \quad I][PH + S]\left[\begin{array}{c|c} W_1 + L_1NL_1^{\mathrm{T}} & 0 \\ \hline 0 & E_2 \end{array}\right]^{-1}. \tag{38}$$

Table I summarizes the optimal observer algorithm derived above. Table II presents the Kalman filter algorithm for the same system, as described by Eqs. (9) and (10). The Kalman algorithm is extracted from the work of Meditch [11] (see also Bryson and Johansen [13] or Bucy [14]) and presented in a form

TABLE I

OPTIMAL OBSERVER ALGORITHM

Observer state estimate update equations

$$\hat{x}_1(t) \equiv y_1(t)$$

$$\hat{v}(t) \equiv 0$$

$$\dot{\hat{x}}_2(t) \equiv F_{21}y_1 + F_{22}\hat{x}_2 + G_2u + K_2(z - H_2\hat{x}_2); \qquad \hat{x}_2(0) = \bar{x}_2$$

Optimal gain matrix

$$K_2 = [Q_{22}^{-1}\, Q_{21} \quad I_{n-m_1}][PH + S]\left[\begin{array}{c|c} W_1 + L_1NL_1^{\mathrm{T}} & 0 \\ \hline 0 & E_2 \end{array}\right]^{-1}$$

Estimation error covariance propagation

$$\dot{P} = \psi(K_2, t)P + P\psi^{\mathrm{T}}(K_2, t) + \Gamma(K_2, t); \qquad P(0) = \text{given},$$

where $\psi(K_2, t)$, $\Gamma(K_2, t)$, S and H are defined in Eqs. (29), (31), (36), and (13), respectively.

TABLE II

KALMAN FILTER ALGORITHM

Estimator state update equations

$$\hat{x}_1(t) = y_1(t) + L_1(t)\hat{v}(t),$$

$$\begin{Bmatrix} \dot{\hat{v}} \\ \dot{\hat{x}}_2 \end{Bmatrix} = \begin{bmatrix} \Phi & 0 \\ M_2 & F_{22} \end{bmatrix} \begin{Bmatrix} \hat{v} \\ \hat{x}_2 \end{Bmatrix} + \begin{Bmatrix} 0 \\ G_2 u + F_{21} y_1 \end{Bmatrix} + \begin{bmatrix} K_1^* \\ K_2^* \end{bmatrix} (z - [H_1, \ H_2] \times \begin{Bmatrix} \hat{v} \\ \hat{x}_2 \end{Bmatrix}); \quad \text{with initial condition,} \begin{Bmatrix} \hat{v}(0) \\ \hat{x}_2(0) \end{Bmatrix} = \begin{Bmatrix} 0 \\ \bar{x}_2 \end{Bmatrix}$$

Kalman gain matrix

$$K^* = \begin{bmatrix} K_1^* \\ K_2^* \end{bmatrix} = (PH^{\mathrm{T}} + S) \begin{bmatrix} W_1 + L_1 N L_1^{\mathrm{T}} & 0 \\ 0 & E_2 \end{bmatrix}^{-1}$$

Estimation error covariance propagation

$$\dot{P} = \psi^*(K^*, t)P + P\psi^{*\mathrm{T}}(K^*, t) + \Gamma^*(K^*, t); \qquad P(0) \text{ given;}$$

where S and H are as defined in Eqs. (36) and (13) and

$$\psi^*(K^*, t) = \begin{bmatrix} \Phi & 0 \\ M_2 & F_{22} \end{bmatrix} - K^* H,$$

$$\Gamma^*(K^*, t) = K^* \left[\begin{array}{c|c} W_1 + L_1 N L_1^T & 0 \\ \hline 0 & E_2 \end{array}\right] K^{*T} - SK^{*T} - K^* S^T + \left[\begin{array}{c|c} N & 0 \\ \hline 0 & W_2 \end{array}\right]$$

---

comparable to that of the observer. Specifically, the covariance equation for $\dot{P}$ in the Kalman algorithm is given in a form that is applicable for any K matrix, including the optimal one K^*. If the optimal K^* is substituted into the right-hand side of the Kalman covariance equation, a considerable simplification results, yielding the more common form of the Kalman filter. This was not done in order to preserve the similarities between the Kalman filter and the newly derived optimal observer algorithm. Several points are worth noting in regards to the two algorithms.

(i) In the Kalman filter, the differential equation for the observer has order $(n - m_1 + q)$ and estimates both the state vector x_2 (with dimension $n - m_1$) and the colored noise vector v (with dimension q). In the observer, however, the estimator differential equation has order $n - m_1$ and estimates only x_2, the estimate of v being always zero. This represents an important savings, particularly in the time-invariant case where only the steady-state filter is to be implemented.

(ii) In both algorithms, integration of an $(n - m_1 + q)$-dimensional matrix Riccati equation is required.

(iii) If, in the Kalman filter algorithm, one replaces the optimal Kalman gain

$$K^* = \left[\begin{array}{c} K_1^* \\ \hline K_2^* \end{array}\right]$$

with a suboptimal gain

$$K^{OBS} = \left[\begin{array}{c} 0 \\ \hline K_2 \end{array}\right] = \left[\begin{array}{c|c} 0 & 0 \\ \hline Q_{22}^{-1} Q_{21} & I \end{array}\right] \left[\begin{array}{c} K_1^* \\ \hline K_2^* \end{array}\right],$$

then the optimal observer algorithm results! This fact makes computer programming of either algorithm very similar.

(iv) The observer optimal gain matrix is not unique. It depends instead on the weighting matrix $Q(t)$ which was transformed by Eq. (24) from the original weighting matrix Q'. For any one system, there are an infinite number of optimal observers corresponding to various choices of the weighting matrix. This is in sharp contrast to a Kalman filter which is unique for a given system and optimal for all positive definite weighting matrices. Note, however, that if the original cost function has no cross weighting between Δx_1 and Δx_2, then $Q_{21}(t) = 0$ and the observer's dependence $Q(t)$ disappears, yeilding a unique optimal observer.

The effect of an additional state transformation $\begin{matrix} H_1^* \\ T \end{matrix}$ as described in comment (ii) of Section II can be determined. If one had an initially defined state vector x^* and some desired cost function $J = E\{\Delta x^{*T} \; Q^* \; \Delta x^*\}$, then the transformation

$$x = \left[\begin{array}{c} H_1^* \\ \hline T \end{array}\right] x^*$$

would lead to $J = E\{\Delta x^T \; Q' \; \Delta x\}$, where

$$Q' = \left[\begin{array}{c} H_1^* \\ \hline T \end{array}\right]^T Q^* \left[\begin{array}{c} H_1^* \\ \hline T \end{array}\right],$$

which would not, in general, have $Q'_{21} = 0$ even if Q^* is diagonal. Thus choice of a diagonal weighting matrix Q^* would generally result in $Q_{21} \neq 0$ and hence an observer solution that does indeed depend on the values in Q^*! To require $Q_{21} = 0$ ($\Rightarrow Q'_{21} = 0$) places restrictions on the allowable choice for Q^*!

(v) The observer does not depend on Q_{11} and in fact does nothing to minimize errors on $\hat{x}_1$ since, by assumption in Eq. (15), $\Delta x_1 = L_1 v$ and cannot be affected by the observer.

(iv) No implicit assumptions have been made concerning the initial value of $\hat{x}_2$. Although no proof is offered, the author feels that stability is assured from arbitrary initial conditions. This was found to be true in all simulated problems including several not given in Section VI.

V. ELIMINATION OF $\dot{y}_1$ FROM FILTER EQUATIONS

Both the Kalman filter and the observer algorithms require inputs involving $\dot{y}_1$ (recall the definition of z_1 in Eq. (12)). This is an undesirable situation since, in practice, it is difficult to form derivatives. This section employs a change of variables first suggested by Stear and Stubberud [14] to eliminate $\dot{y}_1$ from the equations.

Partition the observer gain matrix as

$$K_2 = [K_{21} \,\vdots\, K_{22}]. \qquad (39)$$
$$|\leftarrow m_1 \rightarrow|\leftarrow m_2 \rightarrow|$$

Then use the definition of z in Eq. (12) to substitute into Eq. (21). After a change of variables

$$\hat{x}_2 = \xi_2 + K_{21} y_1, \qquad (40)$$

the result can be written as

$$\dot{\xi}_2 = A\xi_2 + Bu + K'_{21}y_1 + K_{22}y_2, \tag{41}$$

where A is given by Eq. (20) and

$$B = G_2 - K_{21}G_1, \tag{42}$$

$$K'_{21} = F_{21} + AK_{21} - K_{21}F_{11} - K_{22}H_{21} - \dot{K}_{21}. \tag{43}$$

The matrix $\dot{K}_{21}$ can be calculated in terms of $\dot{P}$ by differentiating Eq. (38), and $\dot{P}$ is known from Eq. (35); however, this will not be done here.

Figure 1 depicts the final observer algorithm in block diagram form. A similar substitution into the Kalman algorithm of Table II will eliminate $\dot{y}$ from that algorithm. This is done in a manner wholly analogous to the previous method where the entire state vector is transformed by

$$x = \xi + \begin{bmatrix} K_{11} \\ \hdashline K_{21} \end{bmatrix} y_1.$$

The result, which is similar to Eqs. (41)-(43), is excluded from the present chapter for brevity (see Yocum [5]).

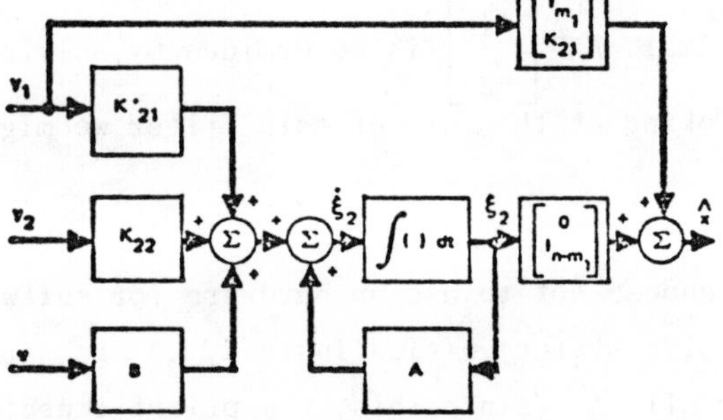

Fig. 1. Observer block diagram.

VI. SUMMARY AND OVERVIEW OF RESULTS

Consider the related but different filtering problem in which the plant and measurement are described by

$$\dot{x} = Fx + Mv + w ,$$

$$z = Hx + Lv + \nu ,$$

where w and ν are white-noise process and v a colored-noise process satisfying

$$\dot{v} = \Phi v + n$$

and n is also a white-noise process. This problem has the essential elements of the problem in Section II after the initial change of variables (i.e., Eqs. (9) and (10)). The Kalman filter solution would combine x and v into a single higher-dimensional state vector and estimate this state vector with a filter of the form:

$$\dot{\hat{v}} = \Phi\hat{v} \qquad + K_1(z - H\hat{x} - L\hat{v}), \tag{44}$$

$$\dot{\hat{x}} = M\hat{v} + F\hat{x} + K_2(z - H\hat{x} - L\hat{v}), \tag{45}$$

where the optimal $K \triangleq \begin{bmatrix} K_1 \\ K_2 \end{bmatrix}$ is determined by solving a Riccati equation. Looking at the form of this filter we might reason as follows:

(i) We choose not to use up hardware (or software) estimating the colored-noise states $v(t)$. Instead, we are satisfied to say $\hat{v}(t) \equiv 0$ for all t (since this an a priori expected value) and thereby avoid solving the first of the two estimator equations.

(ii) By substituting $\hat{v}(t) \equiv 0$ into the second of the two estimator equations we obtain

$$\dot{\hat{x}} = F\hat{x} + K_2(z - H\hat{x}).$$

(iii) Now, however, what is the optimal K_2? At least two choices are apparent:

1. Solve for $K = \begin{bmatrix} K_1 \\ K_2 \end{bmatrix}$ using the usual Kalman algorithm. Then disregard K_1 since it is not needed and use the value for K_2.

2. Noting that $\hat{v}(t) = 0$ is equivalent to assuming that $v(0) = 0$ and solving Eq. (44) for $v(t)$ with $K_1 = 0$, we might go back to the Kalman filter Riccati equation and replace the optimal K with a suboptimal one $K_{\mathrm{OBS}} = \begin{bmatrix} 0 \\ K_2 \end{bmatrix}$; otherwise, use the identical algorithm.

In essence, what has been shown in Section III is that if the cost function is taken to be

$$J = (x - \hat{x})^{\mathrm{T}} Q(x - \hat{x}),$$

where Q is positive definite, then the second approach outlined above is optimum and minimizes J. A slightly more general cost function was considered in which

$$J = \begin{Bmatrix} v - \hat{v} \\ x - \hat{x} \end{Bmatrix}^{\mathrm{T}} \begin{bmatrix} Q_{11} & Q_{21}^{\mathrm{T}} \\ Q_{21} & Q_{22} \end{bmatrix} \begin{Bmatrix} v - \hat{v} \\ x - \hat{x} \end{Bmatrix},$$

and in this case, the expression for the optimum K_2 is a function of $Q_{22}^{-1} Q_{21}$.

VII. EXAMPLE

In this section the observer algorithm is applied to a practical problem of feedback control system design. The example considered is a sixth-order, linear, time-invariant system with a single measurement ($n = 2$, $q = 4$, $m = 1$). Observers of order 1, 2, 4, and 5 are constructed and compared to a true Kalman filter solution of order 5. The comparison includes both estimator performance from a prescribed set of initial conditions and closed loop performance using the steady-state observer in conjunction with a prescribed set of feedback gains.

A. Sytstm Model for Example

The physical system underlying this example is a dual spin spacecraft such as depicted in Fig. 2. Under ideal conditions, the spin axis remains fixed in space and the attitude of the despin platform is specified by the azimuth angle θ. The simplest, but still meaningful, model for the dynamics is

$$\dot{\theta} = \omega, \qquad \dot{\omega} = \frac{1}{I} u + \frac{1}{I} T$$

in which ω is the platform azimuth rate in arcseconds per second, u a despin control torque to be supplied by a feedback loop (in foot-pounds), I the azimuth inertia of platform

$$I = 0.00017 \ \frac{\text{ft-lb}}{\text{arcsec/sec}^2} \ (= 35 \text{ slug ft}^2),$$

and T the friction torque noise (in foot-pounds).

The friction torque noise is generated primarily by the constantly rotating bearing which connects the despun platform to the rotor. It is modeled as a third-order colored-noise

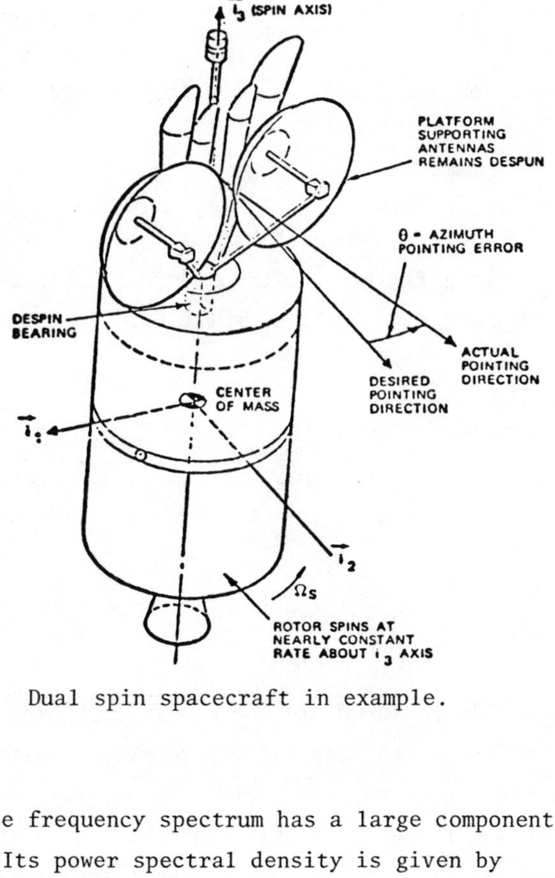

Fig. 2. Dual spin spacecraft in example.

process whose frequency spectrum has a large component near spin frequency. Its power spectral density is given by

$$\Phi_{TT}(s) = \frac{2T_0^2}{\omega_T} \times \frac{s^2 + 2\zeta_1 \Omega s + \Omega^2}{(s^2 + 2\zeta_2 \Omega s + \Omega^2)(1 + s/\omega_T)}$$

$$\times \frac{s^2 - 2\zeta_1 \Omega s + \Omega^2}{(s^2 - 2\zeta_2 \Omega s + \Omega^2)(1 - s/\omega_T)}, \tag{46}$$

where $T_0 = 0.03162$ ft-lb, Ω is the spin frequency (= 6 rpm), $\zeta_1 = 0.75$, $\zeta_2 = 0.05$, and $\omega_T = 20$ rad/sec is the bandwidth of torque noise.

Finally, an on-board attitude sensor provides a measurement of θ:

$$y = \theta + \theta_{err} \tag{47}$$

in which θ_{err} is a first-order colored noise with rms value $\theta_0 = 0.05$ arcsec and bandwidth $\omega_s = 100$ Hz. Its power spectral density is

$$\Phi_{v_1 v_1}(s) = \frac{2\theta_0^2/\omega_s}{(1 + (s/\omega_s))(1 - (s/\omega_s))} . \tag{48}$$

B. Cases Considered

It is possible to formulate the observer problem in a variety of ways, depending on the complexity permissible in the observer (i.e., the dimension of the observer). At one extreme is the fifth-order Kalman filter. In this case the x_2 state vector is fifth-order comprised of ω, θ_{err} and three other variables used to generate T (θ does not appear as an element in the x_2 state vector since it is a linear combination of θ_{err} and the measurement y_1). The colored-noise vector v is nonexistent ($q = 0$). At the other extreme is the minimal order observer of dimension 1. Here the x_2 state vector contains only ω, while the colored-noise vector is fourth order and comprised of θ_{err} plus three states for generating T. Also, the measurement y_1 is used directly as an estimate for θ, and the first-order observer estimates only ω. Various cases in between these extremes are also possible. In all, four cases were considered as defined in Table III. In all cases, the weighting matrix Q in the cost function was diagonal so that the observer becomes independent of Q.

TABLE III

CASES CONSIDERED IN EXAMPLE

Case Number[a]	States in x_1	States in x_2	States in v	Comment
I	θ	ω	θ_{err} $T,\dot{T},\ddot{T}$	Minimal order observer ($n - m_1 = 1$); estimates only ω
II	θ	ω,θ_{err}	$T,\dot{T},\ddot{T}$	Second order observer ($n - m_1 = 2$); estimates ω and θ_{err}
IV	θ	$\omega,T,\dot{T},\ddot{T}$	θ_{err}	Fourth order observer ($n - m_1 = 4$); estimates $\omega,T,\dot{T}$ and $\ddot{T}$
V	θ	$\omega,\theta_{err},T,\dot{T},\ddot{T}$	None	Kalman filter ($n - m_1 = 5$); estimates $\omega,\theta_{err},T,\dot{T},\ddot{T}$

[a]Note that the case number is equal to the dimension of x_2 and therefore the dimension of the resulting filter.

C. Estimator Performance

The observer algorithm was coded on a GE 265 time-sharing computer and the various cases of Table III were run starting from an identical set of initial conditions on covariances. Figs. 3a-c show a comparison of the resulting time histories for

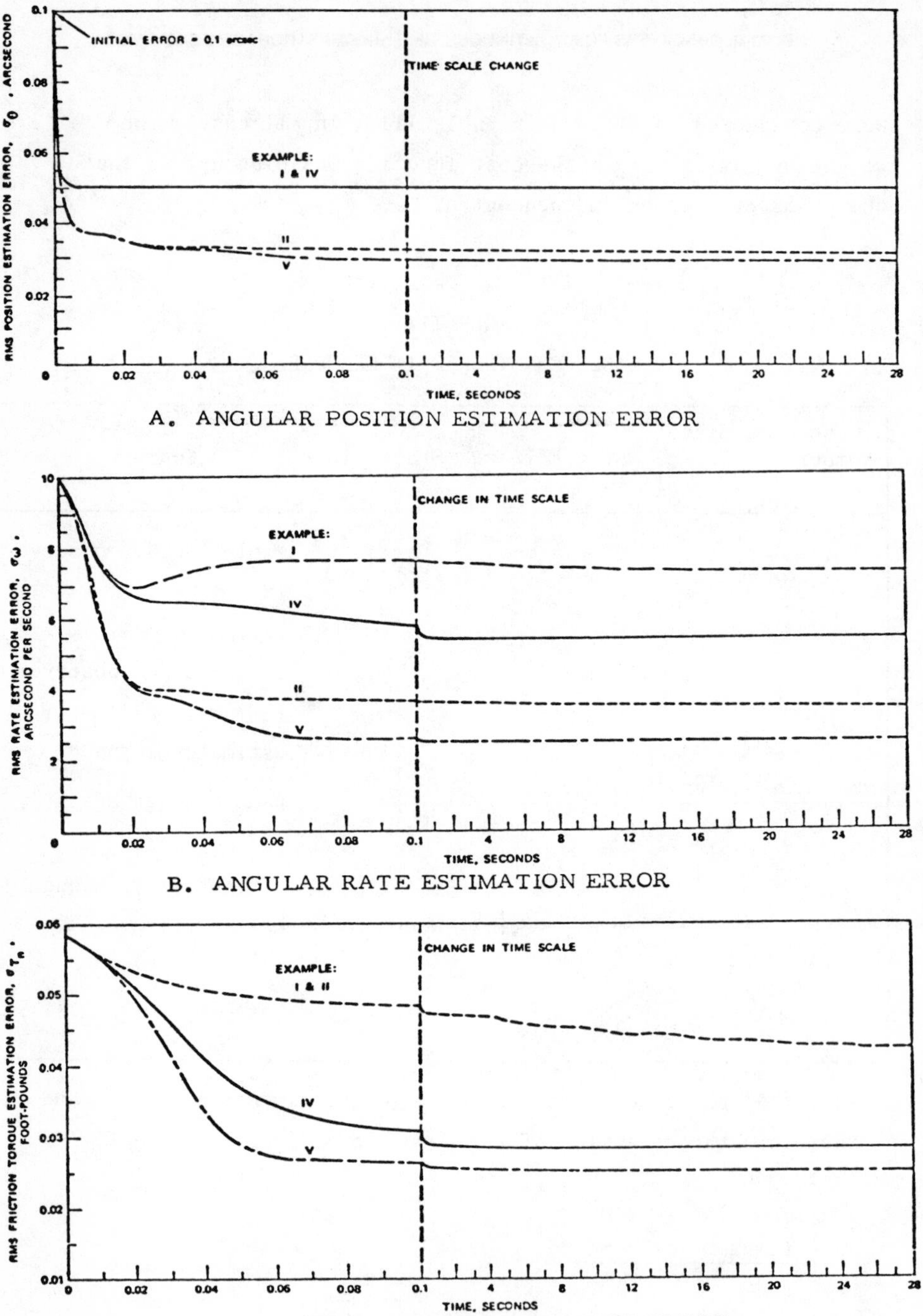

A. ANGULAR POSITION ESTIMATION ERROR

B. ANGULAR RATE ESTIMATION ERROR

C. TORQUE NOISE ESTIMATION ERROR

Fig. 3. Comparison of estimators transient response (Riccati equation solution). (a) Angular position estimation error; (b) angular rate estimation error; (c) torque noise estimation error.

the rms values of the three estimation errors $\Delta\theta_{err}$, $\Delta\omega$, and ΔT (rms estimation error of θ equals that of θ_{err}). Note in Fig. 3a that transient responses for cases I and IV are identical. This is because in these two cases θ_{err} is not estimated by observer. The transient shown is merely the behavior of the defining colored-noise model, in this case given by

$$\dot{P}_{\theta} = -\omega_s P_{\theta} - P_{\theta}\omega_s + 2\theta_0^2\omega_s.$$

A similar statement applies to the torque noise rms response for cases I and II except that the defining colored-noise model is third order (Fig. 3).

Note that settling times for the different order observers are approximately the same. Also, as might be anticipated, an observer estimating a particular state element achieves a lower rms value for that state estimate than an observer that is not estimating that particular state. Compare cases I and IV with II or V in Fig. 3a or cases I and II with IV or V in Fig. 3c. Furthermore, Fig. 3b shows that when different observers estimate the same state element, an increase in the observer dimension results in better estimation. Steady-state rms estimation errors from the figures are tabulated in the second column of Table IV for ease of comparison.

D. Steady-State Observer in a Feedback Path

For the purpose of controlling the platform azimuth angle, a control torque is applied through an electric motor housed within the bearing assembly. The control torque is generated by feeding back the estimated state vector

$$u = -C^{T}\hat{x}, \tag{49}$$

TABLE IV

SUMMARY OF ESTIMATION AND CONTROL RESULTS

Example No.	Steady-State Rms Estimation Error θ (arcsec) ω (arcsec/sec) T_n (ft-lbs)[a]	Open Loop Compensation Filter[b] $U(S)/Y(S)$ $L(\omega) \triangleq (s/\omega) + 1;\quad Q(\Omega,\xi) \triangleq (s^2/\Omega^2) + 2\xi(s/\Omega) + 1$	Filter Eigenvalues (Closed Loop Roots Of Filter) (rad/sec)
I	0.0500[c] 7.082 0.04124[c]	$0.07\ \frac{\text{ft-lb}}{\text{arcsec}}\ \frac{L(11.77)}{L(108.0)}$	$L(30.63)$
II	0.03181 3.347 0.04124[c]	$0.1271\ \frac{\text{ft-lb}}{\text{arcsec}}\ \frac{L(15.68)\ L(628.3)}{Q(157.5,\ 0.7650)}$	$Q(112.0, 0.7300)$
IV	0.0500[c] 5.384 0.02856	$0.08706\ \frac{\text{ft-lb}}{\text{arcsec}}\ \frac{Q(0.6288,0.7501)Q(17.32,0.9652)}{Q(0.6283,0.0500)L(20.00)L(95.76)}$	$Q(0.6283,0.7500)$ $Q(25.59,0.7327)$
V	0.02928 2.502 0.02515	$0.2592\ \frac{\text{ft-lb}}{\text{arcsec}}\ \frac{Q(0.6290,0.7492)Q(23.88,0.9349)L(628.3)}{Q(0.6283,0.0500)L(20.00)Q(136.9,0.7657)}$	$Q(0.6283,0.7500)$ $Q(74.35,0.5106)$ $L(75.44)$

[a]Torque noise state requires three-dimensional state vector for generation or estimation

[b]Units are: ω and Ω in radians per second and ξ, the damping ratio, is dimensionless.

[c]These values are a result solely of noise properties. States so marked are not estimated by observer.

where C^{T} is a 1×6 vector of constant gains and $\hat{x}$ is the 6×1 vector of estimated states from the steady-state observer. For a Kalman filter, all six state have nonzero estimates, whereas one or more of the state estimates are zero for a reduced-order filter. A matrix C was generated as the steady-state solution to an infinite terminal time optimal control problem and has the following values:

$$C_{\hat{\theta}} = 0.252 \frac{\text{ft-lb}}{\text{arcsec}}, \qquad C_{\hat{\omega}} = 0.0131 \frac{\text{ft-lb}}{\text{arcsec/sec}}$$

$$C_{\hat{\theta}_{\text{err}}} = 0, \qquad C_{\hat{T}} = 1.^{\dagger} \tag{50}$$

With the feedback specified, the closed loop control system takes the form in Fig. 4. The "shaping filter" is defined as the filter from measurement input to control torque output and is specified by its Laplace transfer function $-U(S)/Y(S)$.

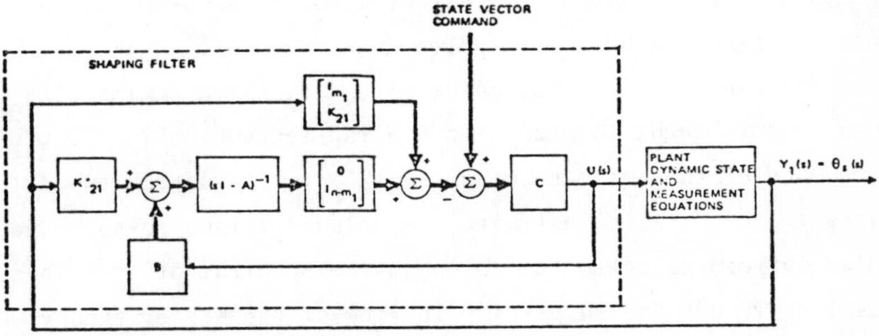

Fig. 4. Feedback control system using observer.

---

†Actually, $\hat{T}$ is generated by a three-dimensional estimator, and three gains are required to specify feedback. Actual gains depend on the way in which the three states are defined, but net effect is such that $\hat{T}$ estimate is fed back to null actual torque noise.

It can be calculated in matrix form by combining the Laplace transfroms of equations (40), (41), and (49):

$$-U(s) = C^{T}\{[sI - (A - BC^{T})]^{-1}(K_{21}' - BC^{T}K_{21}) + K_{21}\}Y_1(s)$$
$$+ C^{T}[sI - (A - BC^{T})]^{-1}K_{22}Y_2(s). \qquad (51)$$

In general, each term in Eq. (51) represents a matrix of transfer functions. In this example, Y_2 is nonexistent and $Y_1(s)$ and $U(s)$ are both scalar quantities. Equation (51) reduces to a scalar transfer function which may be represented in its more usual form as a ratio of polynomials in s, the order of which is just the dimension of the observer $n - m_1$. Table IV lists the shaping filters in factored form for the four cases considered.

Several observations are worthy of note.

(i) All the filters have equal order in numerator and denominator. This is a result of the direct feedthrough of the nonwhite measurement y. Had y been corrupted by white noise instead, it would not have a direct feedthrough into u, and the filters would have more poles than zeros.

(ii) The two filters, cases II and V, which estimate θ_{err}, both contain zeros located at 628.3 rad/sec, which is the break frequency of the noise spectrum for θ_{err}. Also, the two filters, cases IV and V, which estimate the colored torque noise, have poles corresponding to the poles in the spectrum of T. These results are not too surprising in view of the Wiener theory which could have also been used, with considerably more effort, to generate these filters.

(iii) The open-loop transfer function for the control system of Fig. 4 is simply

$$G_{OL}(S) = (1/Is^2) \times (-U(s)/Y(s))$$

Figure 5 shows Bode plots of $G_{OL}(s)$ for the four cases considered. Observe that in all cases, the gain cross-over frequency occurs at nearly peak phase, indicative of what is generally considered a "good" design in the classical control literature. Also, classical

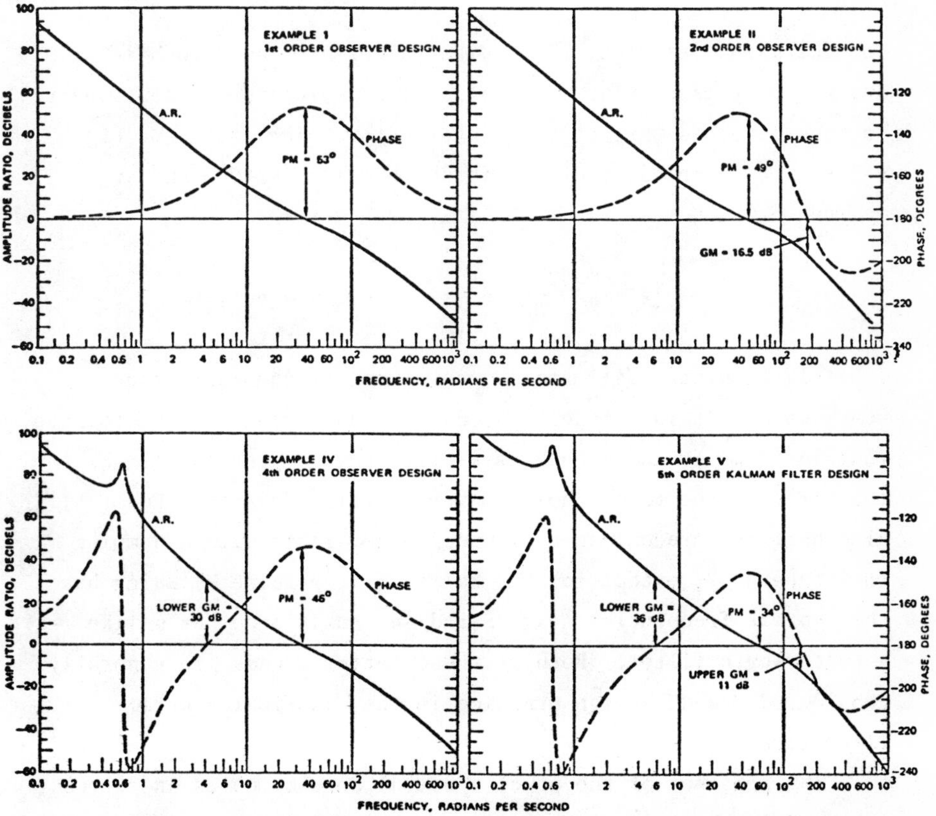

Fig. 5. Comparison of Bode plots for open-loop frequency response curves for control system design. (a) Example I, first-order observer design; (b) Example II, second-order observer design; (c) Example IV, fourth-order observer design; (d) Example V, fifth-order Kalman filter design.

gain and phase stability margins decrease as system complexity increases, but in all cases are within the range generally considered good.

It is well known [2, 7] that the closed-loop roots of the control system divide neatly into those of the system with perfect state feedback plus the eigenvalues of the estimator. The former set of eigenvalues, being independent of the estimator, are the same for all four cases and are given by

$$|sI - (F - C^{\mathrm{T}}G)| = 0.$$

For the gains selected, these eigenvalues are found to be -34.40 and -42.96 rad/sec plus eigenvalue(s) of whichever noise element(s) is considered as part of the state vector. The estimator eigenvalues are independent of the feedback gains in C^{T} and are defined by

$$|sI - A| = 0 = |sI - (F_{12} - K_2H_2)|.$$

Table IV tabulates estimator eigenvalues for the four cases considered. It is difficult to discern any trends except to note that in cases IV and V both estimators have eigenvalues equal to the zeros in the torque noise power spectral density - Eq. (46). Also note the predominance of underdamped roots with a damping ratio near 0.7, except for the Kalman filter (case V) which has three break frequencies near 75 rad/sec positioned not unlike a Butterworth filter! Both of these latter trends are generally considered "good" design practice in the classical control literature.

The presence of the power spectral density poles and zeros in the filter eigenvalues and in the shaping filter has the effect of simply canceling the spectral properties of the corresponding noise when one considers the spectrum of the closed-loop control system output. That is to say, if an observer is designed to estimate a colored noise with spectrum $\Phi_{nn}(s) = F(s)F(-s)$, then the transfer function of the closed-loop control system in response to that noise will contain the factor $1/F(s)$! This result is not altogether surprising and is also predicted by the Wiener theory. Figures 6a and b compare closed-loop frequency responses to sensor noise and torque noise for the four cases considered. Spectrum cancellation is especially evident in Fig. 6b, where cases IV and V, which estimate the colored torque noise, are seen to generate a notch at 0.6283 rad/sec which exactly cancels the peak in the torque noise spectrum. Cases I and II do not estimate T and are seen to have a flat response across the frequency range centered at 0.6283 rad/sec.

A final, general observation is that when an observer estimates a given colored-noise input, the resulting closed-loop system response to that input is lower than for a system that does not estimate the colored noise. This is in agreement with one's intuition and is evident in the low-frequency portion of Fig. 6b and in the high-frequency portion (>628.3 rad/sec) of Fig. 6a.

VIII. CONCLUSIONS

An algorithm, similar in form to the Kalman filter algorithm, has been presented which facilitates design of optimal linear observers of any order between the minimum $n - m_1$ and the maximum order $n - m_1 + q$. The maximum order observer is, in fact, a Kalman filter. Noise sources exciting the system may be white or colored or a combination. The additional states due to colored noise need not increase the observer's dimension, although improved performance results if colored noises are estimated, with a resultant increase in observer dimension. The observers are optimal in the sense that each one minimizes a cost function involving the estimation error covariance. The minimization is

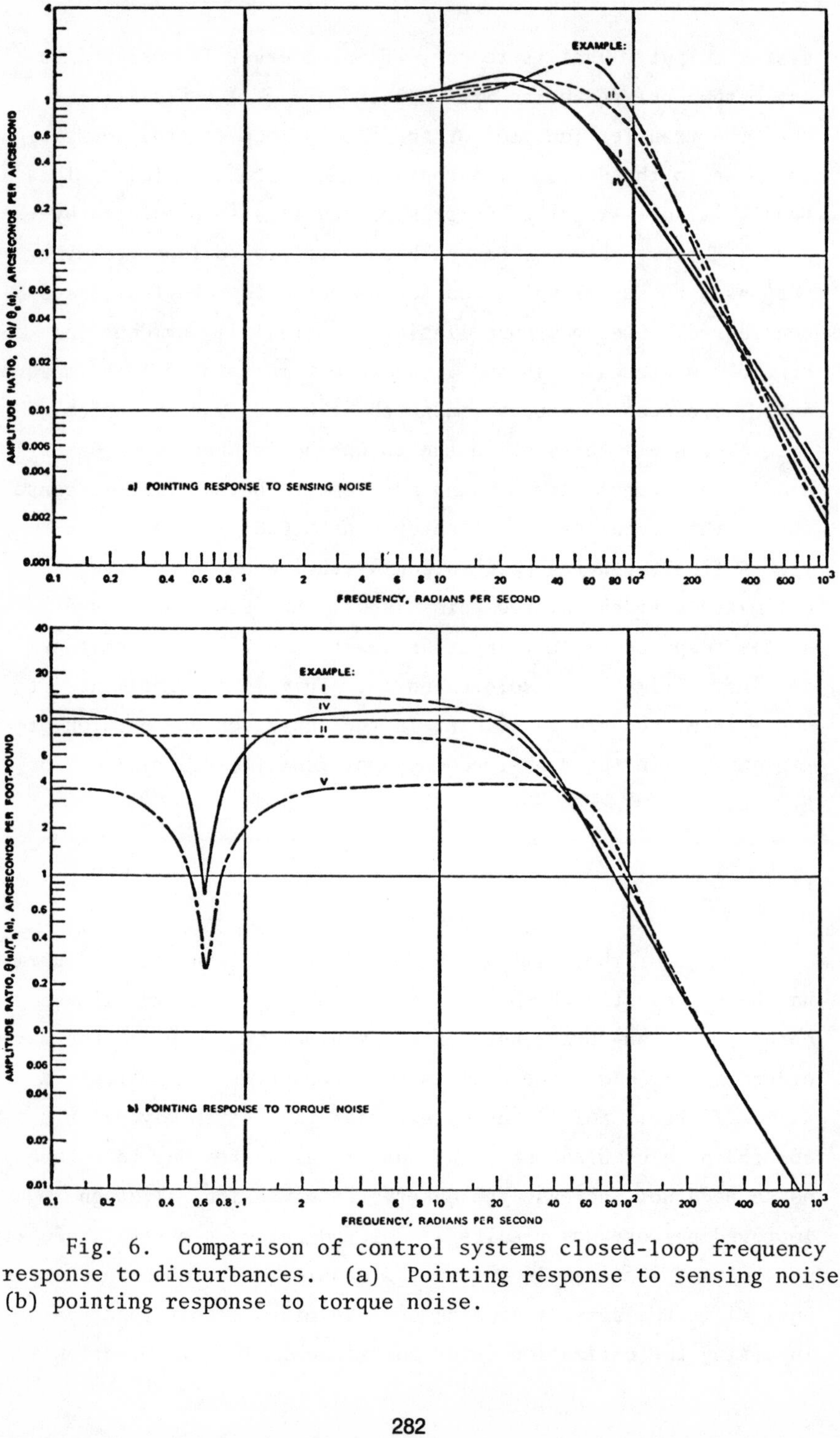

Fig. 6. Comparison of control systems closed-loop frequency response to disturbances. (a) Pointing response to sensing noise; (b) pointing response to torque noise.

subject to dimensionality constraints placed on the observer; and, as is to be expected, larger-order observers achieve better performance, approaching in the limit the Kalman filter's performance.

A disadvantage of the observer algorithm is that it requires solution of matrix Riccati covariance equation equal in dimension to that of the Kalman filter. This will probably restrict applications to cases in which the filter gains can be computed a priori and stored for future use. The chief application of the observer algorithm is foreseen to be in the design of linear time invariant feedback control systems.

A major objection to applying linear optimal stochastic control theory has been that the resulting Kalman filter dimension equals the state dimension and is usually higher than actually necessary for satisfactory control. The observer algorithm, in conjunction with the well-known quadratic cost optimal control problem, eliminates this objection by permitting design of an optimal system which is lower in order than the Kalman filter solution. The lower-order observer design is optimal subject to its dimensionality constraint and its performance is, naturally, not as good as the full Kalman filter solution.

REFERENCES

1. R.E. KALMAN, J. Basic Eng. 82D, 34-45 (1960).
2. D.C. LUENBERGER, IEEE Trans. Military Electronics MIL-8, 74-80 (1964).
3. L.M. NOVAK and C.T. LEONDES, Proc. 4th Asilomar Conf. Circuits Systems, November 1970, 549-552 (1971).
4. L.M. NOVAK, "The Design of an Optimal Observer for Linear Discrete-Time Dynamical Systems". Ph.D. in Eng. Univ. California, Los Angeles, California, June 1971.
5. J.F. YOCUM, "Reduced Order Stochastic Observer Theory for Linear Systems". Ph.D. in Eng. Univ. California, Los

Angeles, California, July 1972.

6. C.T. LEONDES, L.M. NOVAK, and J.F. YOCUM, Proc. 5th Asilomar Conf. Circuits Systems, November 1971, 131-135 (1972).

7. C.T. LEONDES, and L.M. NOVAK, IEEE Trans. Automatic Control, AC-19, 42-46 (1974).

8. E. TSE and M. ATHANS, IEEE Trans. Automatic Control, AC-15, 416-426 (1970).

9. E. TSE and M. ATHANS, Information and Control, 22, 405-434, (1973).

10. D.G. LUENBERGER, "Observers for Multivariable Systems", IEEE Trans. Automatic Control, AC-11, 190-197 (1966).

11. J.S. MEDITCH, "Stochastic Optimal Linear Estimation and Control". McGraw-Hill, New York, New York, 1969.

12. M. ATHANS, Information and Control, 11, 592-606 (1968).

13. E.A. BRYSON and D.E. JOHANSEN, IEEE Trans. Automatic Control, AC-10, 4-10 (1965).

14. E.B. STEAR and A.R. STUBBERUD, Internat. J. Control, 8, 123-130 (1968).

15. R.S. BUCY, "Optimal Filtering for Correlated Noise". Rand Corp. RM 5107 PR, 1966.

OPTIMAL ESTIMATION AND CONTROL OF ELASTIC SPACECRAFT

VICTOR LARSON

Jet Propulsion Laboratory
Pasadena, California

PETER W. LIKINS

School of Engineering and Applied Science
University of California
Los Angeles, California

---

*Research for this chapter was carried out at the Jet Propulsion Laboratory (JPL), California Institute of Technology, under Contract NAS 7-100, sponsored by the National Aeronautics and Space Administration.

I. INTRODUCTION

During the past decade, a very substantial body of literature (see Modi [1] for a partial list) has emerged to permit mathematical modeling and dynamic analysis of complex modern spacecraft. In sharp contrast, applications of modern control theory to nonrigid spacecraft are just beginning to enter the open literature [2-4], in anticipation of the adoption of these methods for flight projects. The spacecraft attitude control community is being forced by the combination of stringent performance requirements and dynamically complex configurations to resort to methods of state estimation and control requiring the use of on-board digital computers, which are now being developed for this purpose [5]. Computer development is now subsiding as the primary obstacle to the practical implementation of modern control theory in spacecraft attitude control, and more subtle limitations of the theory must be confronted and resolved.

A basic difficulty lies in the complexity of the dynamical system required for a high-fidelity model of a modern spacecraft, which typically has dynamically significant flexible appendages, moving parts, system nonlinearities, and various uncertainties. The best available model of the spacecraft, to be called the <u>evaluation model</u>, may then be a nonlinear, stochastic system of very high dimension (involving perhaps as many as 100 states). Because of the practical constraints of attitude control computer size, one cannot incorporate the evaluation model in the state estimator of the spacecraft; one may, however, adopt a linearized and severely truncated model (the <u>estimator model</u>) in its place. With different models in "plant" and "filter" the formal theory of modern control is compromised immediately, with consequences to be determined either empirically or by expansion of present theory [6].

In this chapter, the standard solution [7] of the linear, quadratic, Gaussian problem of optimal estimation and control is required. In order to establish notational conventions, this result is recorded here.

Given the linear equations of state x and measurement z,

$$\dot{x} = Fx + Gu + \Gamma w, \tag{1}$$

$$z = Hx + v, \tag{2}$$

where v and w are white, Gaussian noise terms, and given a fixed-time, quadratic cost functional as the expected value

$$J = E\left[\frac{1}{2}\, x^T(T) S_T x(T) + \frac{1}{2}\int_0^T (x^T A x + u^T B u)\, dt\right], \tag{3}$$

the optimal control vector u minimizing J for control over the interval $0 \leq t \leq T$ is given by

$$u^* = -C\hat{x}, \tag{4}$$

where

$$C = B^{-1} G^T S \tag{5}$$

with S satisfying

$$\dot{S} = -SF - F^T S + C^T BC - A, \tag{6}$$

$$S(T) = S_T, \tag{7}$$

and where the estimated state $\hat{x}$ is obtained from

$$\dot{\hat{x}} = F\hat{x} + Gu^* + K(z - H\hat{x}) \tag{8}$$

with

$$K = PH^T R^{-1} \tag{9}$$

and with P satisfying

$$\dot{P} = FP + PF^T - KRK^T + \Gamma Q \Gamma^T. \tag{10}$$

The initial value of P is obtained from the initial covariance of the error estimate, since

$$E[(x - \hat{x})(x - \hat{x})^T] \triangleq P \tag{11a}$$

and the quantities Q and R are covariances obtained from the autocorrelations

$$E[w(t) - \bar{w}(t)][w(\tau) - \bar{w}(\tau)]^T \triangleq Q(t)\delta(t - \tau) \tag{11b}$$

and

$$E[v(t) - \bar{v}(t)][v(\tau) - \bar{v}(\tau)]^T \triangleq R(t)\delta(t - \tau) \tag{11c}$$

where an overbar denotes mean value, and $\delta(t - \tau)$ is the Dirac function. The designer must select S_T, A, and B.

As indicated, the essential obstacle to the straightforward application of this theory to spacecraft attitude control lies in the difficulty in characterizing the system as in Eqs. (1) and (2) with sufficient fidelity, while at the same time keeping the state dimension within flight computer capacity. In this chapter this problem is confronted for a specific vehicle, and associated obstacles to practical implementation are identified and resolved.

II. SPACECRAFT EVALUATION MODEL

The spacecraft selected as the vehicle for the examination of the problems of modern control is a Solar Electric Spacecraft which was extensively developed at JPL as a paper study extending over a period of several years [8, 9]. This spacecraft was designed for electric propulsion during deep space operations, requiring large, lightweight solar panels, with natural periods of vibration in early designs approaching one minute in duration.

Figure 1 illustrates the essential features of the spacecraft. Attitude control torques are applied by differential gimbaling of six constant-force ion propulsion thrusters, as described by Marsh [8]. Solar panels have an adjustable orientation relative to the primary body, so that a system of hybrid coordinates has been employed, consisting of discrete attitude angles for the central body and the motors and distributed (or "modal") coordinates for the solar panels. The solar panel coordinates correspond to mode shapes and natural frequencies associated with panel boundary conditions which permit no rotation of the central body, while permitting translation restrained only by spacecraft mass.

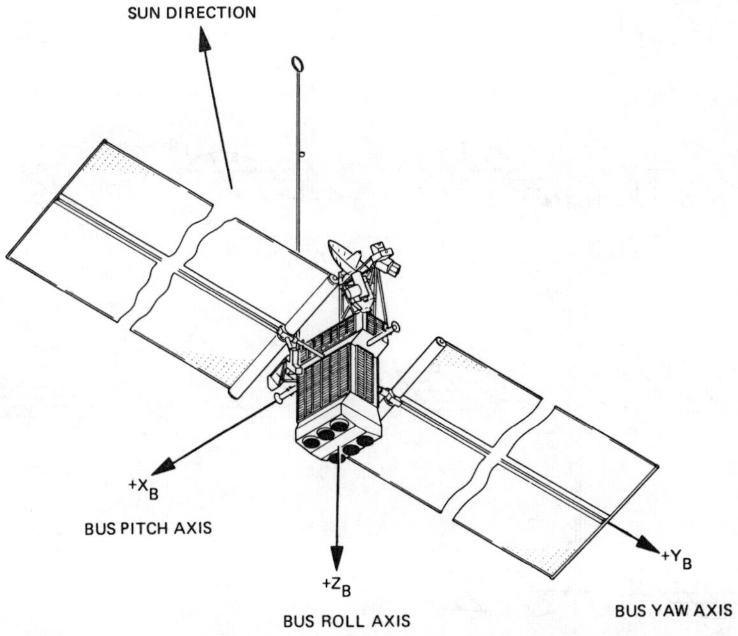

Fig. 1. Pictorial sketch of SEP spacecraft.

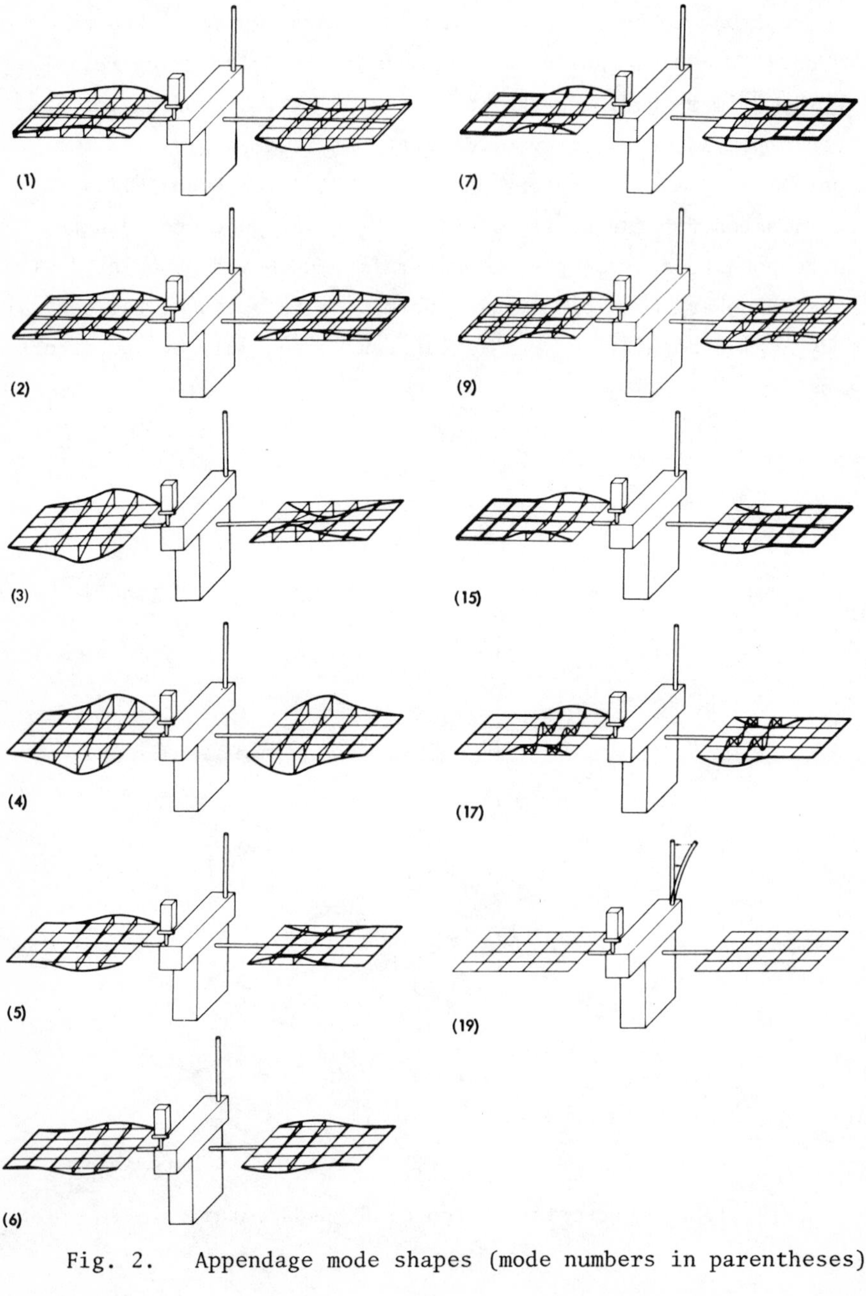

Fig. 2. Appendage mode shapes (mode numbers in parentheses).

Figure 2 illustrates the mode shapes of the six panel modes of lowest frequency and five additional modes of particular interest in this study. The modal data were obtained from a finite-element model of the panels.† Numerical values of relevant system parameters may be found in Table I, which shows the following: mode number (according to ascending frequency), natural frequency, quantities δ_{jx}, δ_{jy}, δ_{jz}, which measure the coupling of mode j with central body rotations about axes x, y, and z, respectively, and quantities Δ_{jx}, Δ_{jy}, and Δ_{jz}, which measure the coupling of mode j with translation of the central body in the directions of axes x, y, and z, respectively.

TABLE I

APPENDAGE HYBRID MODAL DATA

Mode Number j	Frequency σ_j (Hz)	Coupling Scalars					
		δ_{jx}	δ_{jy}	δ_{jz}	Δ_{jx}	Δ_{jy}	Δ_{jz}
1	0.070	172.16	0	0	0	0	0
2	0.074	-0.24	0.46	0	0	0	-3.36
3	0.09632	0	0.01	20.99	0	0	0
4	0.09635	0	9.6	-0.02	0.14	0	0
5	0.09929	0	0	-177.86	0	0	0
6	0.1043	0	0.89	0.23	-3.16	0	0
7	0.10642	-4.95	0	0	0	0	0
9	0.156	-41.33	0	0	0	0	0
15	0.2281	2.48	0	0	0	0	0
17	0.23648	-1.63	0	0	0	0	0
19	0.23953	12.22	0	2.15	0	0.51	0

Appropriate combinations of equations from the work of Marsh [8, 9] produce scalar equations of motion which can be

† Structural analysis was completed by Dr. Ronald Ross of JPL.

written in a natural matrix form of mixed order as

$$I\ddot{\theta} - \delta^T\ddot{\eta} = b_R u + b_R w_R, \tag{12}$$

$$\ddot{\eta} + 2\zeta\sigma\dot{\eta} + \sigma^2\eta - \delta\ddot{\theta} = b_V u + w_V, \tag{13}$$

$$\tau\dot{V}_S + V_S = K_S\theta + K_S\phi^S\eta + w_S, \tag{14}$$

$$V = V_S + v, \tag{15}$$

where

$$\theta \triangleq [\theta_x \ \ \theta_y \ \ \theta_z]^T; \quad u \triangleq [\alpha_x \ \ \alpha_y \ \ \gamma_x \ \ \gamma_y]^T; \quad \eta \triangleq [\eta_1 \ \ \eta_2 \ \cdots \ \eta_N]^T;$$

$$w_R \triangleq [w_x \ \ w_y \ \ w_z]^T; \quad w_V \triangleq [w_1 \ \ w_2 \ \cdots \ w_N]^T; \quad w_S \triangleq [w_{Sx} \ \ w_{Sy} \ \ w_{Sz}]^T;$$

$$v \triangleq [v_x \ \ v_y \ \ v_z]^T; \quad V \triangleq [V_x \ \ V_y \ \ V_z]^T; \quad V_S \triangleq [V_{Sx} \ \ V_{Sy} \ \ V_{Sz}]^T;$$

and the remaining symbols are constant matrices. When the equations are given this form, it becomes apparent that they describe an entire class of spacecraft attitude regulation problems. (Compare Eqs. (12) and (13) with Eqs. (288) and (289) of the paper by Likins [10].) For the purposes of this chapter, actuator dynamics will be ignored, and gimbal angles will be treated as control variables.

In these hybrid coordinate equations, the central body attitude angles are called θ_x, θ_y, and θ_z, and the appendage modal vibration coordinates are called $\eta_1, \cdots, \eta_N$, for N modes. The gimbal angle combinations called α_x, α_y, γ_x, and γ_y are the control variables, to be specified as functions of the state. Sensor measurements V_x, V_y, and V_z are noisy representations of sensor state variables V_{Sx}, V_{xy}, and V_{Sz}.

The term $K_s \phi^s \eta$ in the sensor equation shows any direct contribution of deformations to sensor orientation. Additive noise terms, represented by w_x, w_y, w_z, w_1, $\cdots$, w_N, w_{sx}, w_{sy}, w_{sz}, v_x, v_y, and v_z, are all treated as white and Gaussian. All other symbols in the equations represent constant parameters, which will be identified explicitly only as needed for physical interpretation. The scalars δ_{jx}, δ_{jy}, δ_{jz} in Table I comprise the jth row of δ in Eq. (13), and the terms Δ_{jx}, Δ_{jy}, and Δ_{jz} appear in b_v (see Marsh [8]).

It is important to realize that Eqs. (12) - (15) represent an imperfect model of the physical system. It may be necessary to modify the established theory (see the Introduction) before spacecraft attitude estimation and optimal control systems can be designed with the confidence that they will function in the presence of these errors; progress in this direction has been reported [11-13] and is continuing (see, for example, Skelton and Likins [14]). Nonetheless, in the present chapter the game is to be played by the old rules, and the established linear quadratic Gaussian theory is to be used to obtain a design for state estimation and control based on a linear model of dimension small enough for flight computation. Then an evaluation is to be made of system performance with a simulation corresponding to the block diagram in Fig. 3, employing the more elaborate evaluation model in the physical system or "plant" block while the estimator model is used in the state estimator of "filter" block. By introducing some of the indicated modeling error adjustments in the evaluation model, and exercising the simulation, one can explore the adequacy of present theory and provide a measure of the need for extensions of the kind advanced in References 11-14.

Equations (12)-(15) must be recast in the state variable form of Eqs. (1) and (2) before the general results in the Introduction can be applied. The equivalence of Eqs. (1) and (2) with Eqs. (12)-(15) is obtained by setting

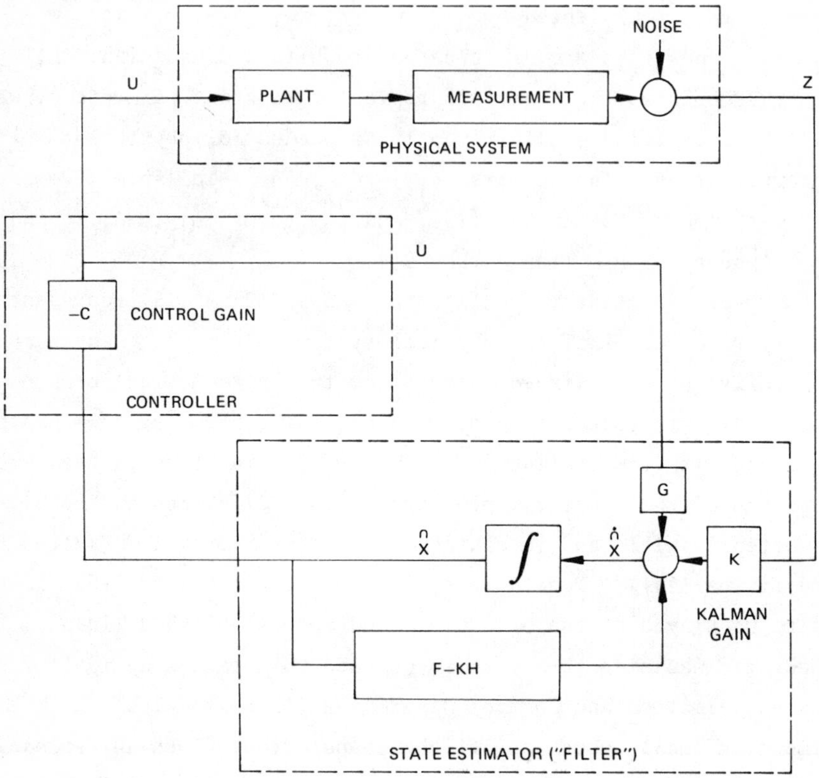

Fig. 3. Block diagram: optimal estimation and control.

$$x \triangleq [\theta^T \quad \dot{\theta}^T \quad \eta^T \quad \dot{\eta}^T \quad V_s^T]^T, \tag{16a}$$

$$u \triangleq [\alpha_x \quad \alpha_y \quad \gamma_x \quad \gamma_y]^T, \tag{16b}$$

$$w \triangleq [0 \quad w_R^T b_R^T \quad 0 \quad w_V^T \quad w_s^T]^T \tag{16c}$$

such that

$$\Gamma = \begin{bmatrix} U & 0 & 0 & 0 & 0 \\ 0 & I & 0 & -\delta^T & 0 \\ 0 & 0 & U & 0 & 0 \\ 0 & -\delta & 0 & U & 0 \\ 0 & 0 & 0 & 0 & \tau \end{bmatrix}^{-1} , \tag{17}$$

$$F = \Gamma \begin{bmatrix} 0 & U & 0 & 0 & 0 \\ 0 & 0 & 0 & 0 & 0 \\ 0 & 0 & 0 & U & 0 \\ 0 & 0 & -\sigma^2 & -2\zeta\sigma & 0 \\ K_s & 0 & K_s\phi^s & 0 & -U \end{bmatrix}, \tag{18}$$

$$G = \Gamma[0 \quad b_R^T \quad 0 \quad b_v^T \quad 0]^T, \tag{19}$$

$$H = [0 \quad 0 \quad 0 \quad 0 \quad U]. \tag{20}$$

Here U is a unit matrix, and all other symbols are drawn from Eqs. (12)-(15).

Equations (1) and (2) with the specific interpretations given by Eqs. (16)-(20) provide a basis for optimal estimator and controller design, as established by Eqs. (3)-(11). Such a design would require numerical values of system parameters, for plant and sensor noise characteristics (Q and R, respectively), and for the initial covariance of the error estimate $P(0)$. Finally, the design would depend critically on the choice of control interval T and the weighting matrices S_T, A, and B in the cost functional J [see Eq. (3)].

In the paper by Ohkami and Likins [4] the rank tests are applied to hybrid coordinate spacecraft attitude equations to obtain explicit algebraic criteria, which in the special case with actuators and sensors for three orthogonal axes attached to the primary rigid body and distinct appendage frequencies, reduce simply to

$$\delta^i \delta^{i^T} > 0, \qquad i = 1, \cdots, N, \tag{21}$$

where δ^i is the ith column of the matrix δ appearing in Eq. (17). In the more general case, the interpretation of controllability and observability criteria is facilitated by transformation to (uncoupled) vehicle normal mode coordinates [15]. Note from Table I that for this system all appendage frequencies are distinct (although some are close together) and $\delta^i \delta^{i^T} > 0$ for all i (although this number is much smaller for some modes than for others). The fact that $\delta_{ix}^2 = 0$ for some modes indicates that these modes become uncontrollable if the only control capability is a torque about the x-axis.

Even with all considerations of general theory and comprehensive spacecraft modeling resolved, there remain critical problems of implementation that are largely judgmental in nature. Such problems are perhaps best examined in the context of a particular problem, as developed in the following sections.

III. SPACECRAFT ESTIMATOR MODEL

Constraints of flight computer size require that the mathematical model of the spacecraft employed in the estimator be smaller in size and simpler in structure than would normally be appropriate for the spacecraft evaluation model presented in the previous section. With this constraint in mind, we shall develop in this section nothing larger than a sixth-order, linear model, permitting single-axis simulation of a vehicle with a flexible appendage having two modes of vibration. For the exposition to follow, we limit ourselves further to a fourth-order estimator model, allowing only one mode of vibration. We shall focus on control about the x-axis, which experience shows to be the most troublesome one for the solar electric propulsion spacecraft. Moreover we ignore all noise sources except actuators and sensors (for which we have some empirical data), and suppress the numerically small terms Δ_{jy} $(j = 1, \cdots, N)$ which describe coupling between appendage modal vibrations and primary body translational accelerations (see Table I). Finally we ignore the small influence of appendage deformations on sensor signals (treating the sensors as though they were attached to the primary rigid body, whereas in fact they are on the appendages near the primary body). Each of these simplifications is incorporated in the estimator model only; the evaluation model retains full generality.

With these restrictions, the state equations to be used in constructing the estimator appear again as Eqs. (1) and (2), but now for the hybrid coordinate model we have a fourth-order system with scalar elements as

$$x \triangleq \begin{pmatrix} \theta_x \\ \dot{\theta}_x \\ \eta \\ \dot{\eta} \end{pmatrix}; \qquad w \triangleq w_R; \qquad u \triangleq \alpha_x, \tag{22a}$$

$$F = \frac{1}{I - \delta^2} \begin{bmatrix} 0 & I - \delta^2 & 0 & 0 \\ 0 & 0 & -\delta\sigma^2 & -2\zeta\sigma\delta \\ 0 & 0 & 0 & I - \delta^2 \\ 0 & 0 & -I\sigma^2 & -2I\zeta\sigma \end{bmatrix}, \tag{22b}$$

$$G = \frac{-\sum_{i=1}^{6} F_i n_i L}{I - \delta^2} \begin{pmatrix} 0 \\ 1 \\ 0 \\ \delta \end{pmatrix}; \qquad \Gamma = G, \tag{22c}$$

$$H = [K_{sx} \quad 0 \quad 0 \quad 0]. \tag{22d}$$

Before attempting the computations required by Eqs. (4)-(10) for solution of the optimal estimation and control problem, one should examine the questions of controllability and observability. Since for this model both sensors and actuators are on the primary body, and for the single mode considered here $\delta^2 \neq 0$, both complete controllability and complete observability are assured by the results of Ohkami and Likins [4].

The next step in the control system design is the assignment of values to system parameters. For the fourth-order estimator model, the fixed parameters for the solar electric spacecraft are nominally as follows: $I = 33{,}353$ slug-ft^2; $\delta = 172.16$; $\sigma = 0.070$ Hertz; $\zeta = 0.005$; $\sum_{i=1}^{6} F_i n_i L = 1.12$ ft-lb; $K_{sx} = 300$ V/rad for plant, 1 rad/rad for estimator; $Q = 1 \times 10^{-6}$ rad^2 sec; $R = 2.1025 \times 10^{-8}$ rad^2 sec.

The symbol I is the moment of inertia about the x-axis through the system mass center, which can be calculated or measured prior to flight with an accuracy of a few percent, tempered primarily by in-flight changes due to thermal distortions, consumption of expendables, elastic vibrations and other moving parts. The symbol δ is a measure of the coupling between primary body rotation about the x-axis and vibration in the chosen mode (see δ_{jx} in Table I); this quantity depends on appendage mode shape and mass distribution, and can be calculated reliably for the lower modes to within perhaps 10%. The symbol σ is the natural frequency of vibration of the chosen appendage mode for selected boundary conditions, and is known for the lower frequencies to within a few percent. The modal damping ratio ζ is usually not at all well known, and could be in error by an order of magnitude. For the fourth-order estimator model, we include appendage mode No. 1 from Table I; for the sixth-order model, we include modes Nos. 1 and 9, choosing the latter because δ_{9x} is relatively large. The combination $\sum_{i=1}^{6} F_i n_i L$ is a measure of the torque capacity of the six ion engines, each of which is presumed in the first baseline design to have a constant thrust of $F_i = 30$ mlb, $i = 1, \cdots, 6$. The symbols $n_1, \cdots, n_6$ are binary measures of the participation of the six engines in the control task, and L is an effective "lever arm". (A second baseline design involving 12 engines of 150 mlb thrust is later introduced for comparison.) The sensor gain K_{sx} is taken arbitrarily as unity, so that the sensor measures the attitude angle θ in radians.

The covariance of the measurement noise R has been obtained from data on a Canopus tracker tested at JPL, and the covariance of the plant noise is based on an estimated uncertainty in assigning motor gimbal angles.

The major design decisions are the choices of the weighting matrices A and B in Eq. (3), and the related choice of control interval T and terminal state penalty matrix S_T if T is finite. For the given spacecraft, control during cruise is a regulatory task, for which an infinite control interval is most appropriate, but commanded turns might be required in finite time. For the calculations which follow a typical finite value is assigned to T, and steady-state results are used to obtain the constant solutions approached as T goes to infinity. The matrix B (a scalar in this application) is chosen as unity without loss of generality (beyond the assumption of a nonzero penalty for control action). The terminal state penalty matrix S_T influences only the transient solution, which is of secondary importance here; for the hybrid coordinate model S_T is taken as null except for ones on the main diagonal of the upper left 2 by 2 partition. The critical design decision is the choice of the matrix A in the cost functional.

The rational selection of the elements of A is much more

straightforward when one thinks in terms of the hybrid coordinate formulation of the problem, since then the penalties on primary rigid body rotation states (θ and $\dot{\theta}$) are explicitly distinguished from penalties on appendage deformation states (η and $\dot{\eta}$). Since for this spacecraft the mission requirements are expressed wholly in terms of θ and $\dot{\theta}$, the weighting matrix A in the hybrid coordinate case is null except for A_{11} and A_{22}. If instead of a fourth- or sixth-order estimator model we had a second-order (rigid-body) model, we could solve Eqs. (5) and (6) literally [16] for C in terms of A_{11}, A_{22}, and B and use Eq. (4) at time zero to obtain C also in terms of initial (estimated) state and initial permissible gimbal angle, thereby obtaining guidance in the selection of elements of A which for anticipated initial conditions would not cause violation of the unstated but ubiquitous gimbal angle constraints. For the fourth-order and sixth-order problems at hand, we have chosen to consider nonzero initial conditions on θ only, and to calculate A_{11} as above, using the second-order estimator formula as given by Larson [16]

$$C_{11} = \left(\frac{A_{11}}{B}\right)^{1/2} \tag{23a}$$

together with

$$u^*(0) = -C_{11}\theta(0) \tag{23b}$$

to obtain

$$A_{11} = BC_{11}^2 = B\left[\frac{u^*(0)}{\theta(0)}\right]^2. \tag{23c}$$

For the prescribed maximum gimbal angle ($u_{max} = 25°$) and the anticipated initial disturbance (chosen as the "deadband" value of θ, which is $0.25°$), we find (with $B = 1$) that $A_{11} = 10^4$ would be necessary to satisfy the implicit constraints for the second-order model; as a matter of judgment, we make this same choice for the fourth- and sixth-order system we are designing. The choice of A_{22} would be crucial for a "fast" system, but numerical experiments (limited to the single engine control case) indicate for this rather quiescent control system very little sensitivity to variations in A_{22} ranging from 0 to 10^6. For most computer runs, we have therefore assumed $A_{22} = A_{11} = 10^4$ for the hybrid coordinate formulation of the problem.

With all numerical values fixed at the nominal values as indicated, numerical computations provide the results shown in Figs. 4 and 5. Figure 4 shows time histories of control system and estimator gains, for the hybrid coordinate case. Although for these calculations a finite control interval was selected as $T = 600$ sec, the plots show clearly that transients are brief and steady-state (constant) values of C and K must be only slightly suboptimal for even modest control intervals (while being optimal for infinite T). Figures 5a and 5c show $\theta(t)$ and $u(t)$ respectively, as produced by the constant-gain controller and estimator, with the estimator model used also for the "plant model" of the physical system (see Fig. 3).

Because a 12-engine version of the Solar Electric Spacecraft has been proposed, and because it seems clear that appendage vibration problems must increase with engine thrust, we have examined a second baseline design which involves 12 engines with 150 mlb thrust each, rather than 6 engines with 30 mlb thrust. This increases G and Γ in Eq. (22c) by a factor of 10, and correspondingly increases $\Gamma^T Q \Gamma$ in Eq. (10). Changes in the steady-state values of C and K are significant, as indicated in Table II for the hybrid coordinate case. We must expect that the control problem will be more difficult for the second (high-thrust) baseline design.

TABLE II

DEPENDENCE OF STEADY-STATE GAINS ON THRUST

Gains	Engines	
	6 × 30 mP = 0.18 lb	12 × 150 mP = 1.8 lb
C_{11}	-100	-100
C_{12}	-2422	-734.5
C_{13}	0.50	0.36
C_{14}	12.3	3.22
K_{11}	2.16×10^{-2}	7.02×10^{-2}
K_{21}	2.34×10^{-4}	2.46×10^{-3}
K_{31}	4.83×10^{-3}	4.67×10^{-1}
K_{41}	5.56×10^{-5}	2.75×10^{-2}

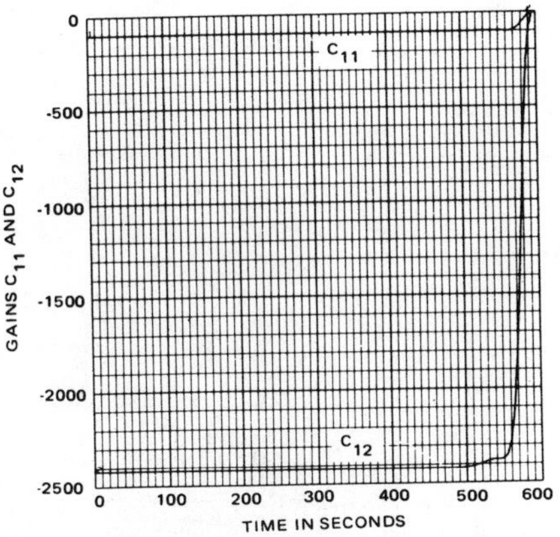

(a) Hybrid Control Gains C_{11} and C_{12}

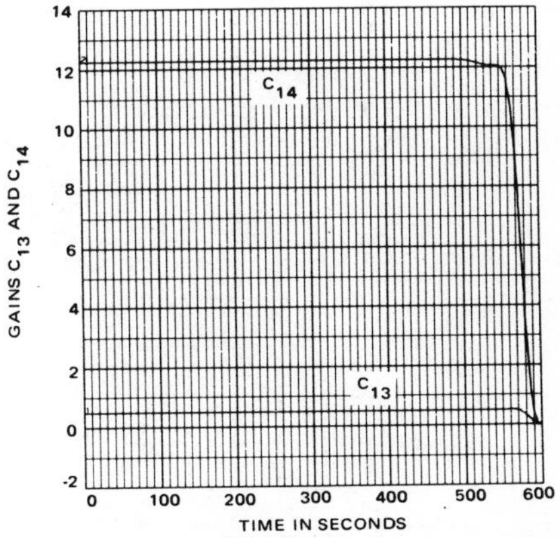

(b) Hybrid Control Gains C_{13} and C_{14}

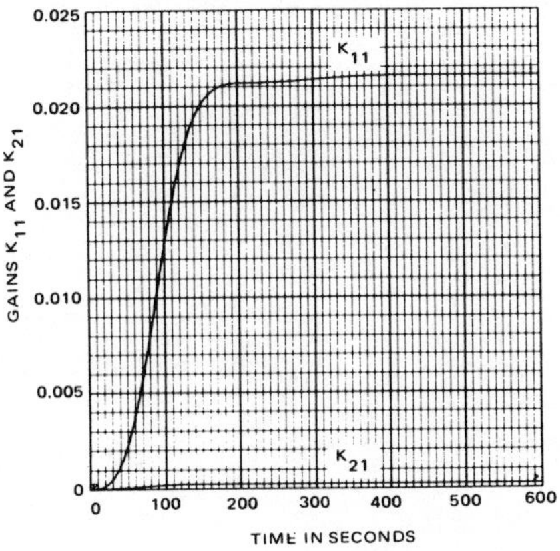

(c) Hybrid Filter Gains K_{11} and K_{21}

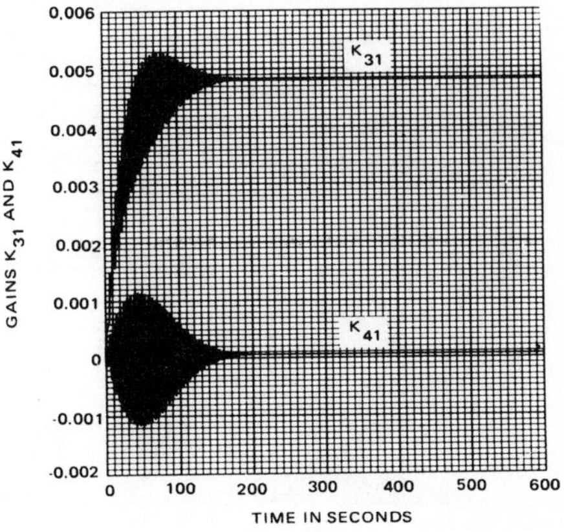

(d) Hybrid Filter Gains K_{31} and K_{41}

Fig. 4. Control and estimator gains for single-axis case: hybrid coordinates. (a) Hybrid control gains C_{11} and C_{12}; (b) C_{13} and C_{14}. (c) Hybrid filter gains K_{11} and K_{21}; (d) K_{31} and K_{41}.

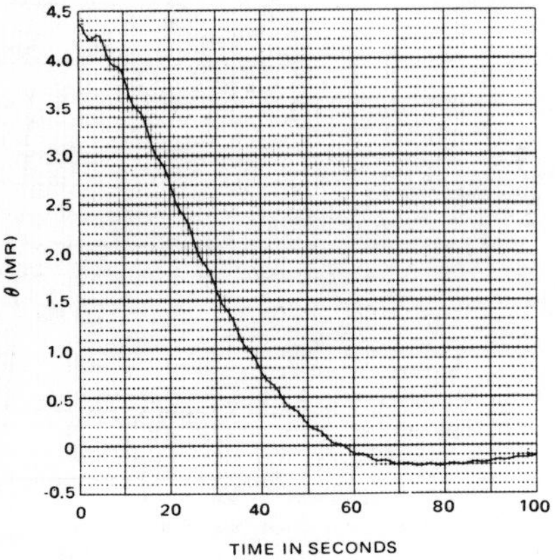

(a) θ History

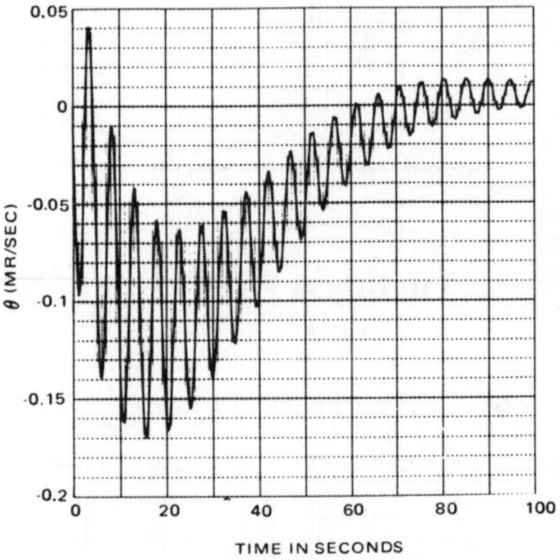

(b) θ History

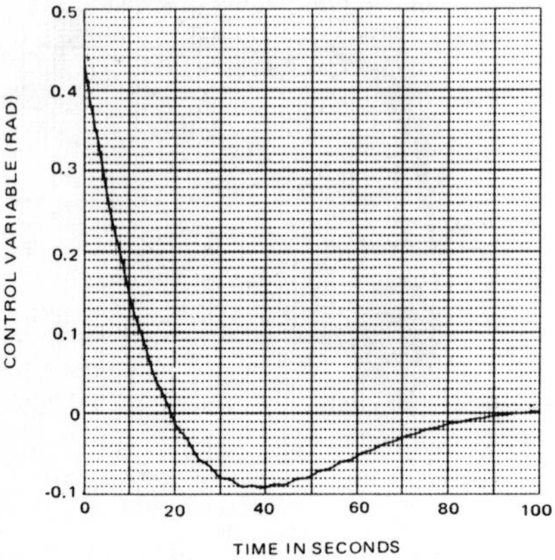

(c) Control Variable History

Fig. 5. Attitude angle, rate, and control gimbal angle time histories for constant gain system. (a) θ history; (b) $\dot{\theta}$ history; (c) control variable history.

IV. SYSTEM EVALUATION

In the case of the solar electric spacecraft control system evaluation, the best that we can do for preflight evaluation is to substitute for the physical system (see Fig. 3) our evaluation model, as represented by Eqs. (12)-(15), with the control system incorporating a state estimator based on a model of only second-, or fourth-, or sixth-order. Because we cannot be sure that the evaluation model is correct, we should explore the influence of selected parameter variations and coordinate truncations on the system behavior.

The first evaluation model to be considered here is a fifth-order model, which differs from the indicated fourth-order estimator model only in the introduction of sensor dynamics [see Eq. (14)], with $\tau_{sx} = 0.05$ sec and $K_{sx} = 300$ V/rad. This model was used for a comparative evaluation of estimator models of second- and fourth-orders for both the nominal thrust and the high-thrust designs (see Figs. 6 and 7).

In particular, it should be noted from Fig. 6 that with a fifth-order evaluation model the change from second- to fourth-order estimator model serves primarily to reduce the ripple in the attitude angle signal; this improvement is most evident in the comparison of rate signals in Figs. 6b and 6d, which show that a peak-to-peak residual of about 2.5×10^{-5} rad/sec at $t = 100$ sec for the second-order estimator is cut in half by the fourth-order estimator, with no significant reduction in rate settling or in the corresponding value of the attitude angle, which in either case has a peak-to-peak residual of about 10^{-4} rad at $t = 100$ sec. Thus it would seem that according to the fifth-order evaluation model the nominal-thrust system with a second-order filter could meet all but the most stringent performance requirements.

In Fig. 7 the consequence of a tenfold increase in actuator torque is revealed. Now the appendage vibrations are more excited,

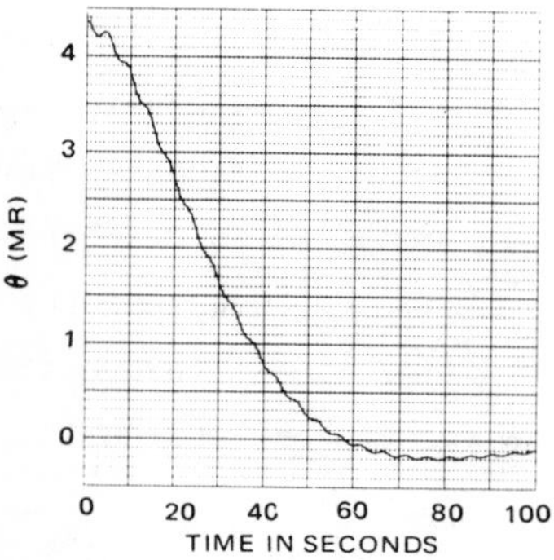

(a) Attitude Angle for 2nd Order Estimator

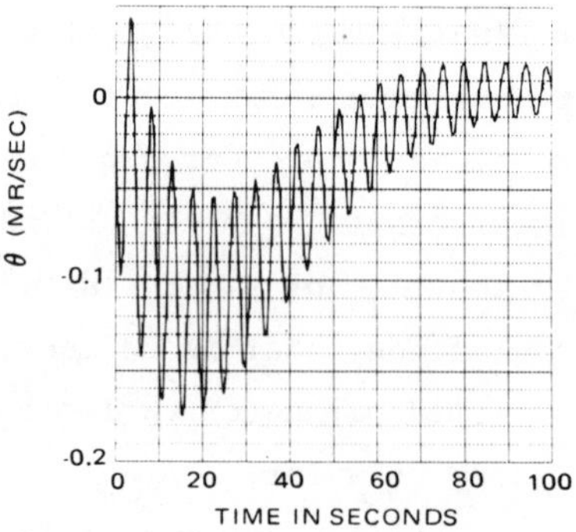

(b) Attitude Rate for 2nd Order Estimator

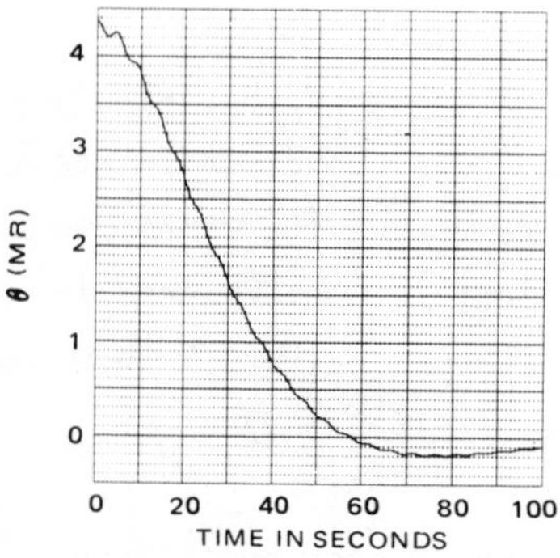

(c) Attitude Angle for 4th Order Estimator

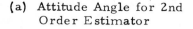

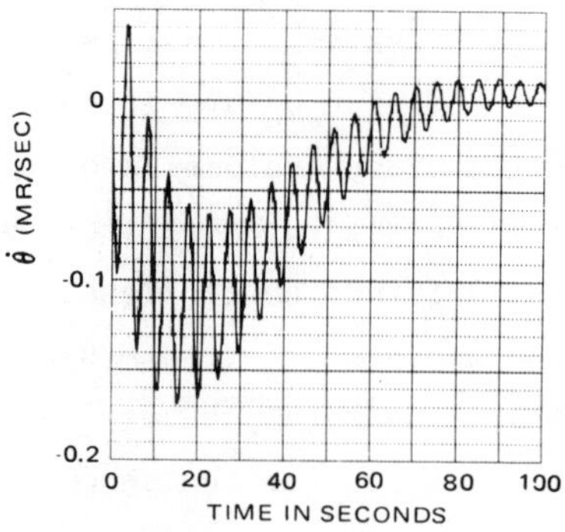

(d) Attitude Rate for 4th Order Estimator

Fig. 6. Comparison of second- and fourth-order estimator models for nominal fifth-order evaluation model. Second-order: (a) attitude angle; (b) attitude rate. Fourth-order: (c) attitude angle; (d) attitude rate.

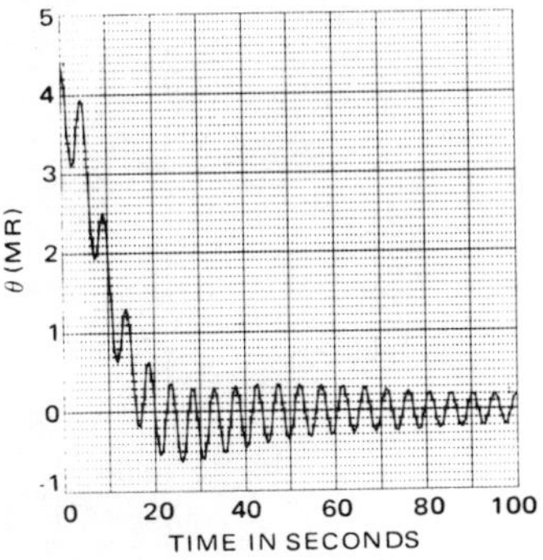

(a) Attitude Angle for 2nd Order Estimator

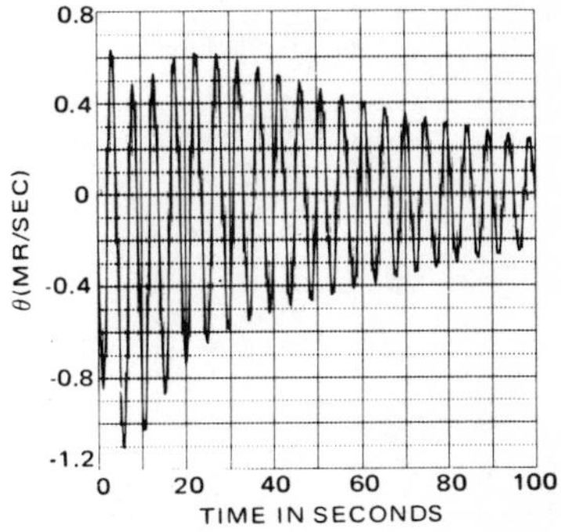

(b) Attitude Rate for 2nd Order Estimator

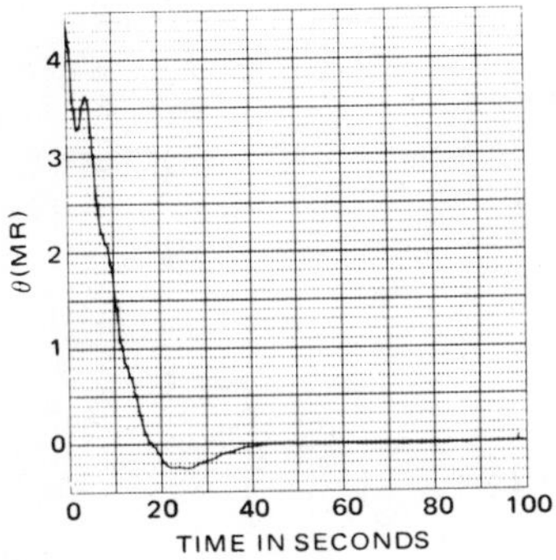

(c) Attitude Angle for 4th Order Estimator

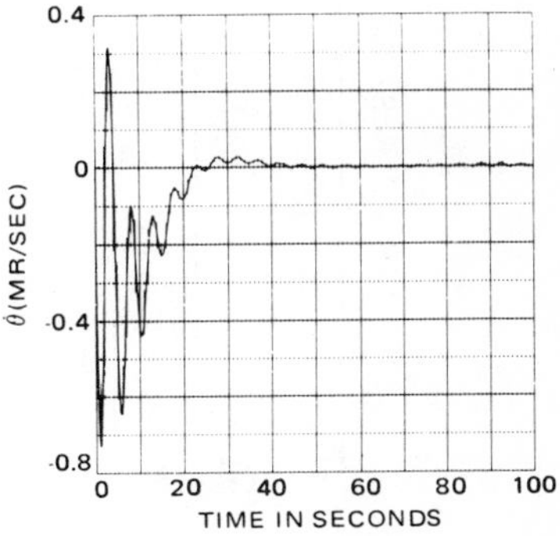

(d) Attitude Rate for 4th Order Estimator

Fig. 7. Comparison of second- and fourth-order evaluation models for high-thrust variation of fifth-order evaluation model. Second-order: (a) attitude angle; (b) attitude rate. Fourth-order: (c) attitude angle; (d) attitude rate.

TABLE III

SUMMARY OF RESIDUAL ATTITUDE ANGLES AND RATES[a]

Case						Peak-to-Peak Attitude and Attitude Rate versus Filter Order					
Plant Order	Parameters					Θ(mr)			$\dot{\Theta}$(mr/sec)		
η_p	N_{ENG}	Thrust (mlb)	ζ	Sensor Noise σ_v (mrad)	R	$\eta_F = 6$	$\eta_F = 4$	$\eta_F = 2$	$\eta_F = 6$	$\eta_F = 4$	$\eta_F = 2$
15	12	150	0.005	0.145	R_N	0.04	0.20	0.20	0.04	0.20	0.20
	12	150	0	0.145	R_N	0.6	1.63		1.26	3.30	
	12	150	0.005	$0.145(10)^{1/2}$	$10 \times R_N$	0.1	0.23		0.08	0.23	
	6	30	0.005	0.145	R_N		0.10	0.10		0.02	0.02
5	12	150	0.005	0.145	R_N		0	0.4		<0.01	0.5
	12	150	0	0.145	R_N		0.03			0.03	
	12	150	0.005	$0.145(10)^{1/2}$	$10 \times R_N$		0			0.02	
	6	30	0.005	0.145	R_N		0.1	0.1		0.12	0.025

[a] Recorded for $t = 100$ sec. Note that a blank entry indicates computation was not made. Nominal values of parameters: $R_N = 2.1025 \times 10^{-8}$, $Q_u = 1 \times 10^{-6}$, $\sigma_u = 1 \times 10^{-3}$; Q_u and σ_u were not varied.

and the inclusion of appendage flexibility in the estimator model is more important. The initial rate settling values are now dramatically improved by going to the larger filter, and the residual peak-to-peak rates at $t = 100$ sec are reduced from 5×10^{-4} rad/sec to well below 10^{-5} rad/sec. Even the residual peak-to-peak attitude errors after 100 sec are reduced, falling from nearly 4×10^{-4} rad to an imperceptible value on the plots (less than 10^{-5} rad). With the more powerful actuators, a shift to the fourth-order estimator may be required by mission objectives, according to the fifth-order evaluation model.

The top half of Table III provides a summary of salient features of Figs. 6 and 7, together with the results of sensitivity studies on structural damping and sensor noise for a fifth-order evaluation model. The bottom half of this table summarizes corresponding results for an evaluation model of higher order, as depicted in more detail in Figs. 8 and 9.

The selection of an evaluation model for final evaluation of a spacecraft control system is a difficult game, with a substantial subjective element. One might argue plausibly that the fifth-order model is big enough for all practical purposes of single-axis control, since from Table I it would appear to any seasoned observer that flexible mode number 1 is clearly dominant, having at once the lowest frequency σ_j and the largest rotational coupling term δ_{jx}. Caution suggests, however, that some consideration be given to the best model available, however large, and for this spacecraft there were many appendage modes at hand. As a compromise for this study we have elected to extend consideration only to a fifteenth-order model, retaining in Eqs. (12)-(15) the single rotation θ_x, six selected coordinates of appendage modal vibration, and the sensor variable V_{sx}. The selected modes are numbered 1, 7, 9, 15, 17, and 19 in Table I, chosen as the modes of lowest frequency which by virtue of δ_{jx} couple with θ_x. In addition to the modal data in Table I, we assumed a nominal damping ratio $\zeta = 0.005$ for all modes, and

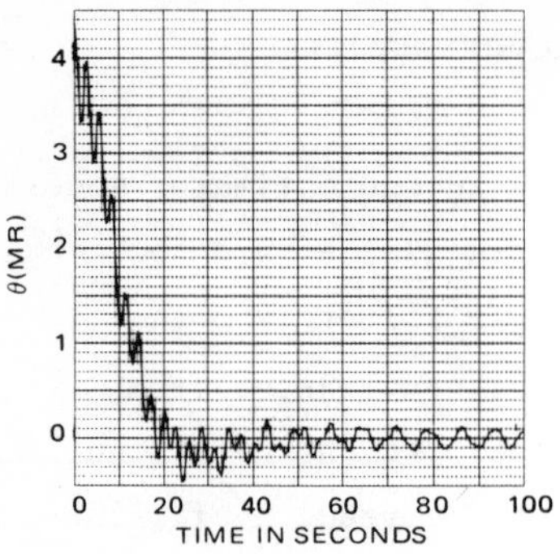

(a) Attitude Angle for 2nd Order Estimator

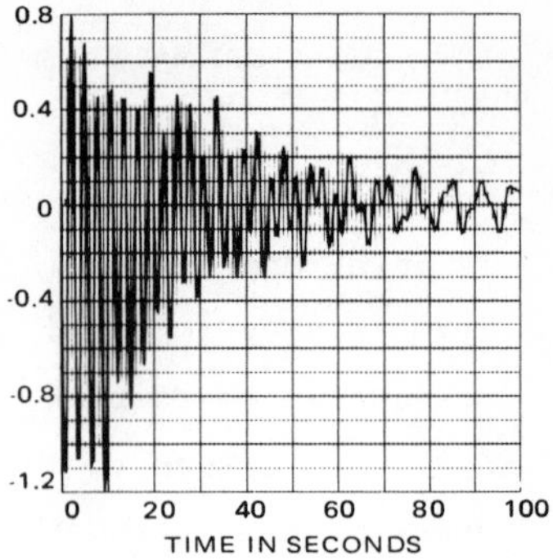

(b) Attitude Rate for 2nd Order Estimator

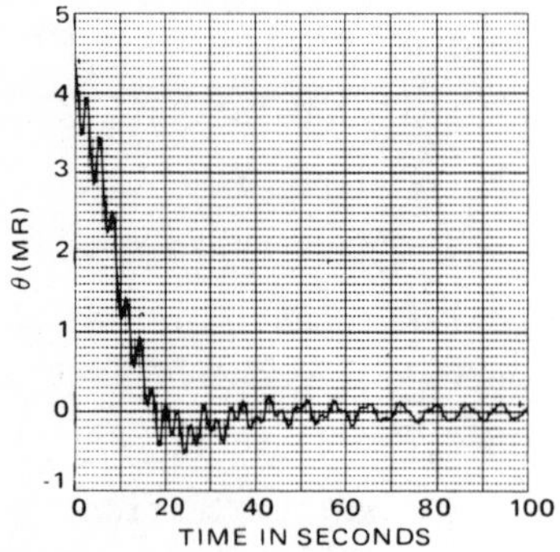

(c) Attitude Angle for 4th Order Estimator

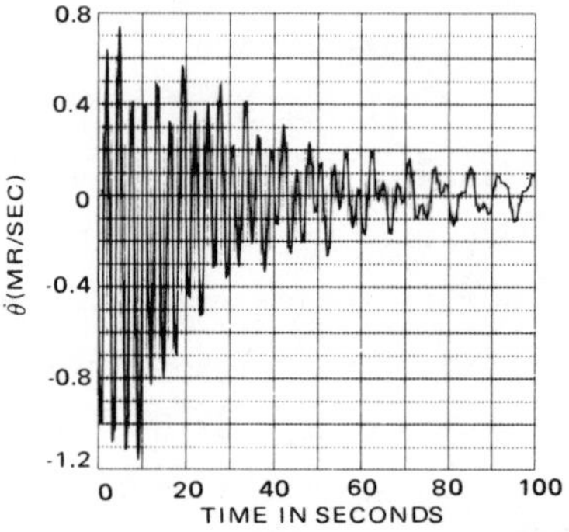

(d) Attitude Rate for 4th Order Estimator

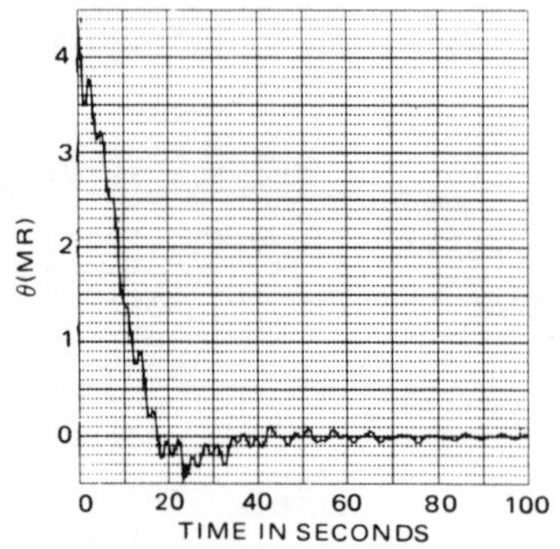

(e) Attitude Angle for 6th Order Estimator

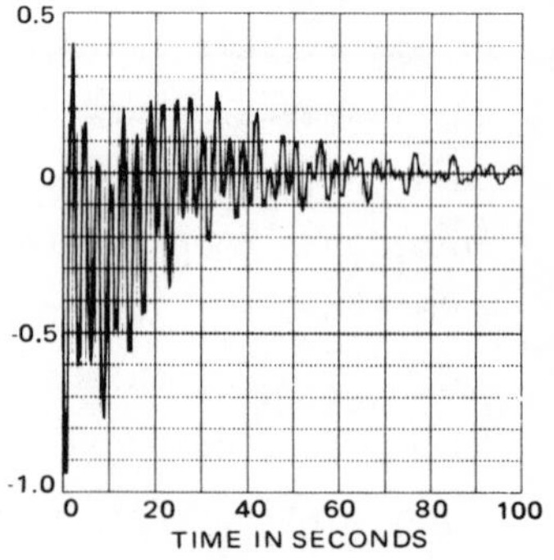

(f) Attitude Rate for 6th Order Estimator

Fig. 8. Comparison of second-, fourth-, and sixth-order filters for fifteenth-order evaluation model; high-thrust controller. Second-order: (a) attitude angle; (b) attitude rate. Fourth-order: (c) attitude angle; (d) attitude rate. Sixth-order: (e) attitude angle; (f) attitude rate.

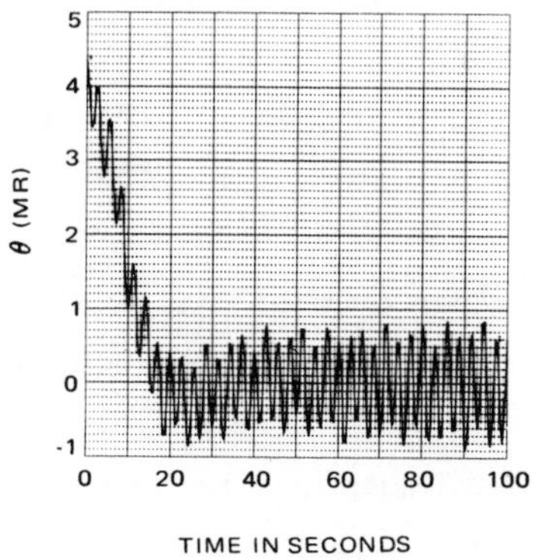

(a) Attitude Angle for 4th Order Estimator

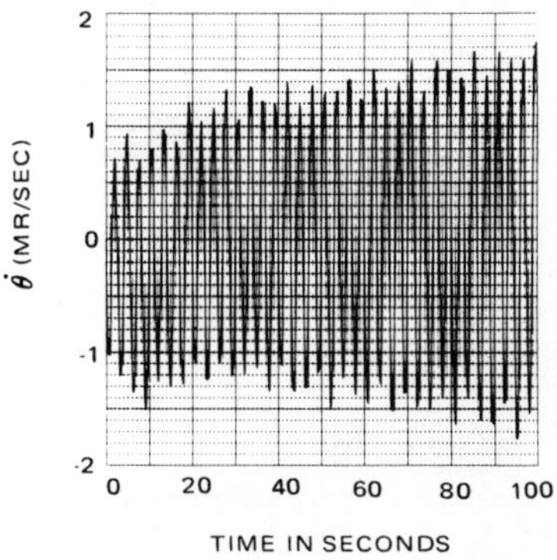

(b) Attitude Rate for 4th Order Estimator

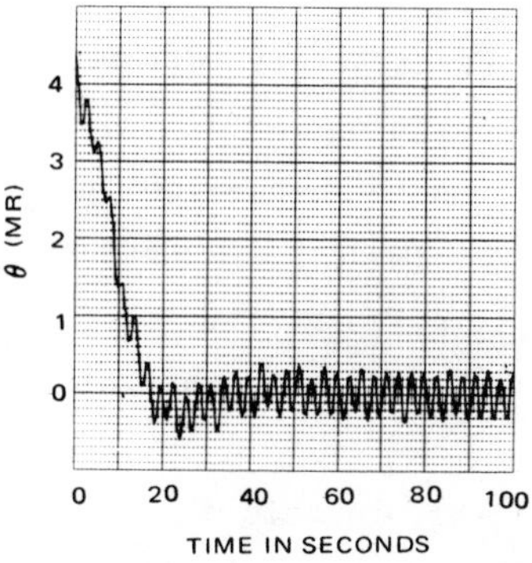

(c) Attitude Rate for 6th Order Estimator

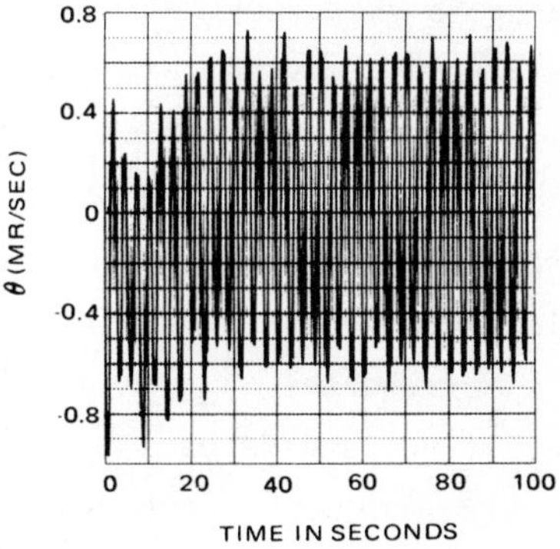

(d) Attitude Rate for 6th Order Estimator

Fig. 9. Variation with no structural damping: high-thrust controller and fifteenth-order evaluation model. Fourth-order: (a) attitude angle; (b) attitude rate. Sixth-order: (c) attitude angle; (d) attitude rate.

from available mode shapes and sensor locations found $\phi^s = 0$.

Figure 8 holds one of the real surprises of this study. This figure is parallel to Fig. 7, with input differing only in the size of the estimator model. We are concerned here not only with the intriguing observation that the second-order estimator does a better job on the fifteenth-order "plant" than on its fifth-order counterpart (compare Figs. 8b and 7b), but also with the absence of this phenomenon with the fourth-order estimator, which does a distinctly better job with the fifth-order plant than with its fifteenth-order counterpart (compare Figs. 8d and 7d). Similar runs were made for the nominal thrust case, with results that are similar in character but less influenced by appendage vibrations, indicating very good performance by most practical standards in every instance. It seems clear that for the nominal thrust case the second-order filter design would suffice, but for the high-thrust design variation the performance of the second-order estimator may be inadequate, and when extension to the fourth-order estimator is accomplished (compare Figs. 8b and 8d) one discovers with chagrin that with the fifteenth-order plant nothing is gained by increasing the estimator order from two to four! Yet it may be that neither gives satisfactory performance. Only by introducing yet another mode into the estimator model (mode number 19 of Table I) is the performance significantly improved, as revealed by Figs. 8e and 8f. (Mode 9 is a natural choice because the coupling term δ_{9x} is next in size to δ_{1x}, and the frequency σ_9 is still not much higher than that of alternative candidates.)

Figure 9, when compared to Figs. 8c-8f, reveals the sensitivity of performance to structural damping, and illustrates also the influence of estimator order on this sensitivity. For this data, and for similar data in Table III, the estimator and controller gains remain those steady-state values calculated for the case of nominal damping ratio ($\zeta = 0.005$), while ζ is zero elsewhere.

Sensitivity studies were also performed on sensor noise, changing evaluation and estimation model noise levels together to ten times the nominal value. Results are summarized in Table III.

V. FREQUENCY DOMAIN ANALYSIS

In order to gain additional insight on the controllers developed in previous sections, a frequency domain analysis was also performed. Bode plots are given below for the flexibility transfer function, the filters, and the system. The transfer functions for the structure and the filters are given by:

$$\text{Structure:} \quad Z(s) = H\Phi_F(s)u(s), \tag{24a}$$

$$\text{Filter:} \quad u(s) = -C\Phi_{\mathcal{F}}(s)KZ(s), \tag{24b}$$

where Φ_F, $\Phi_{\mathcal{F}}$ are the transition matrices for the evaluation model and the filter, respectively (the controller matrix $\mathcal{F} = \bar{F} - \bar{G}C - K\bar{H}$).

Figure 10 shows the Bode plots for the sixth-, fourth-, and second-order filters. Note that the higher-order filters are in essence notch filters which compensate for structural resonances.

Figure 11 provides the Bode plots of the flexibility transfer function of the structure (does not include the rigid-body contribution). This figure clearly shows that the Bode plots for two dominant modes (1, 9 of Fig. 2) are approximately the same as those for six modes. The results for one mode are not nearly as close to those for six modes as those for two modes are. Recall that in the time domain it was much more difficult to rationalize the need for two modes in the filter.

Figure 12 shows the Nichols and Bode plots for a second-order filter with six, two, and one mode (s) used in the evaluation model of the plant. The same conclusion can be drawn from this

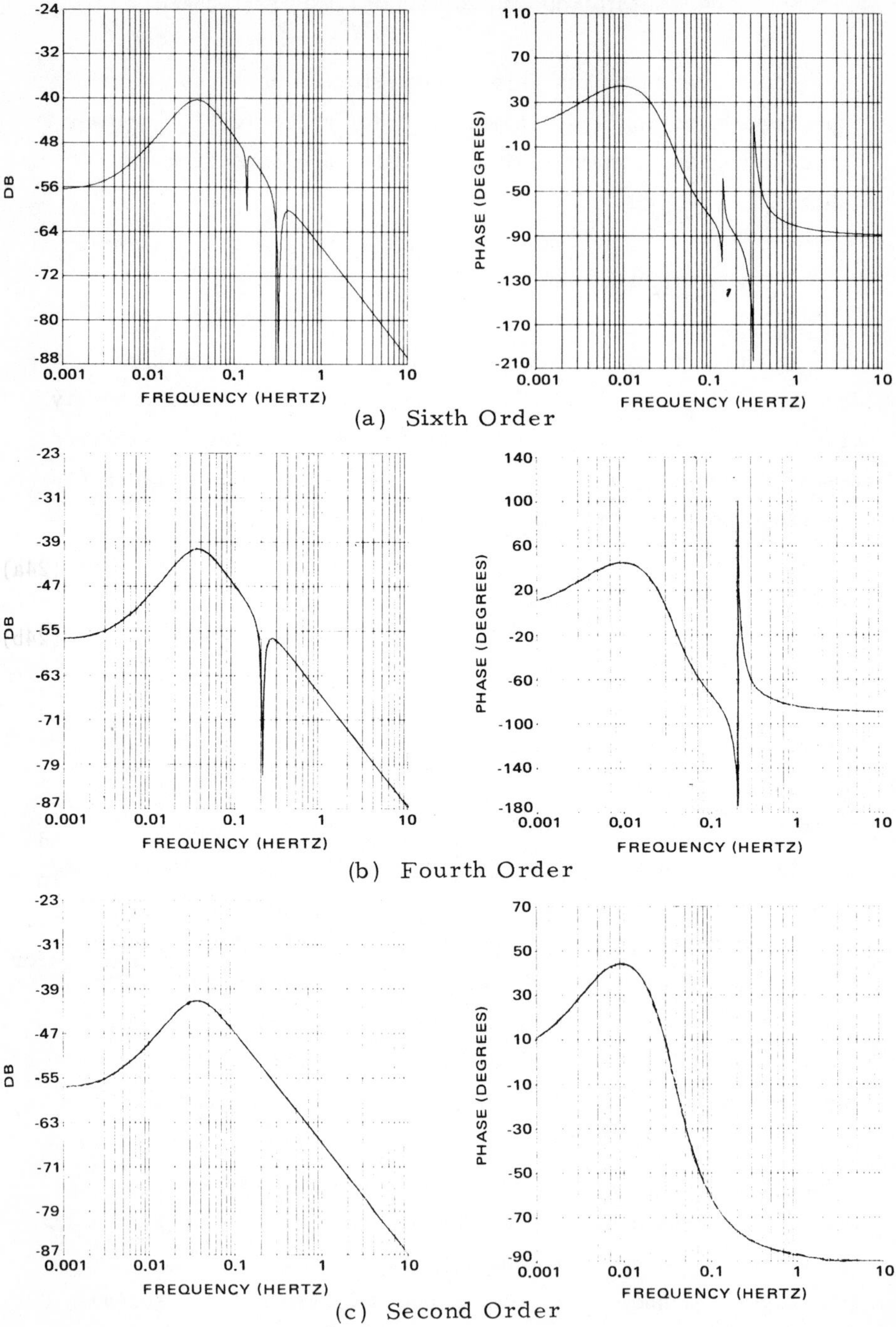

Fig. 10. Open-loop filter Bode plots for (a) Sixth-order; (b) Fourth-order; (c) Second-order.

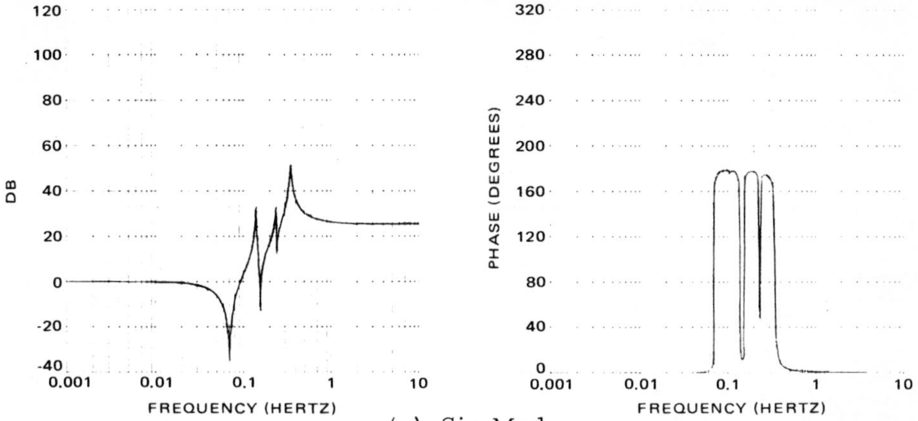

(a) Six Modes

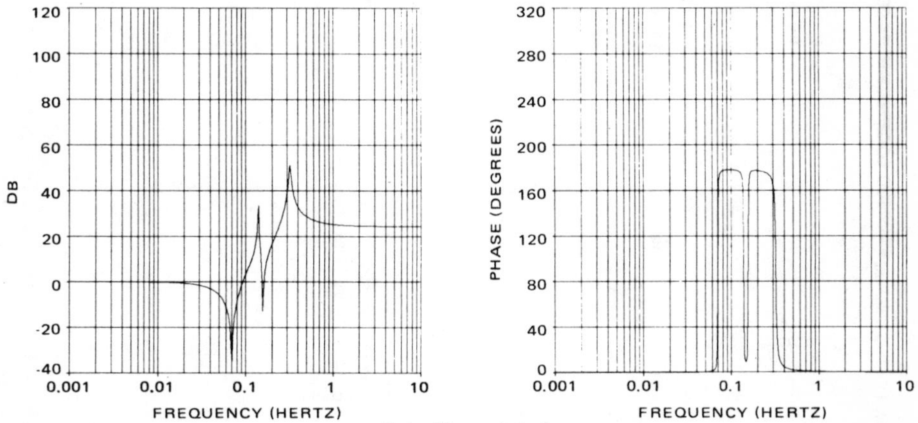

(b) Two Modes

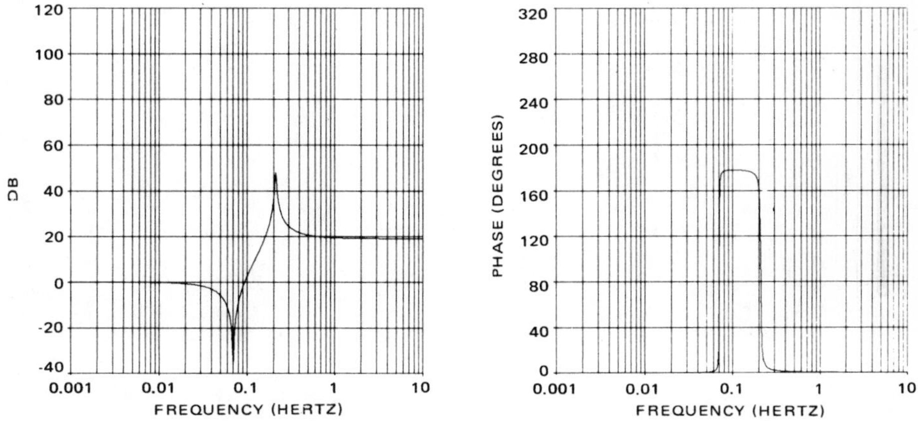

(c) One Mode

Fig. 11. Open-loop Bode plot of flexibility transfer function for (a) six modes; (b) two modes; (c) one mode.

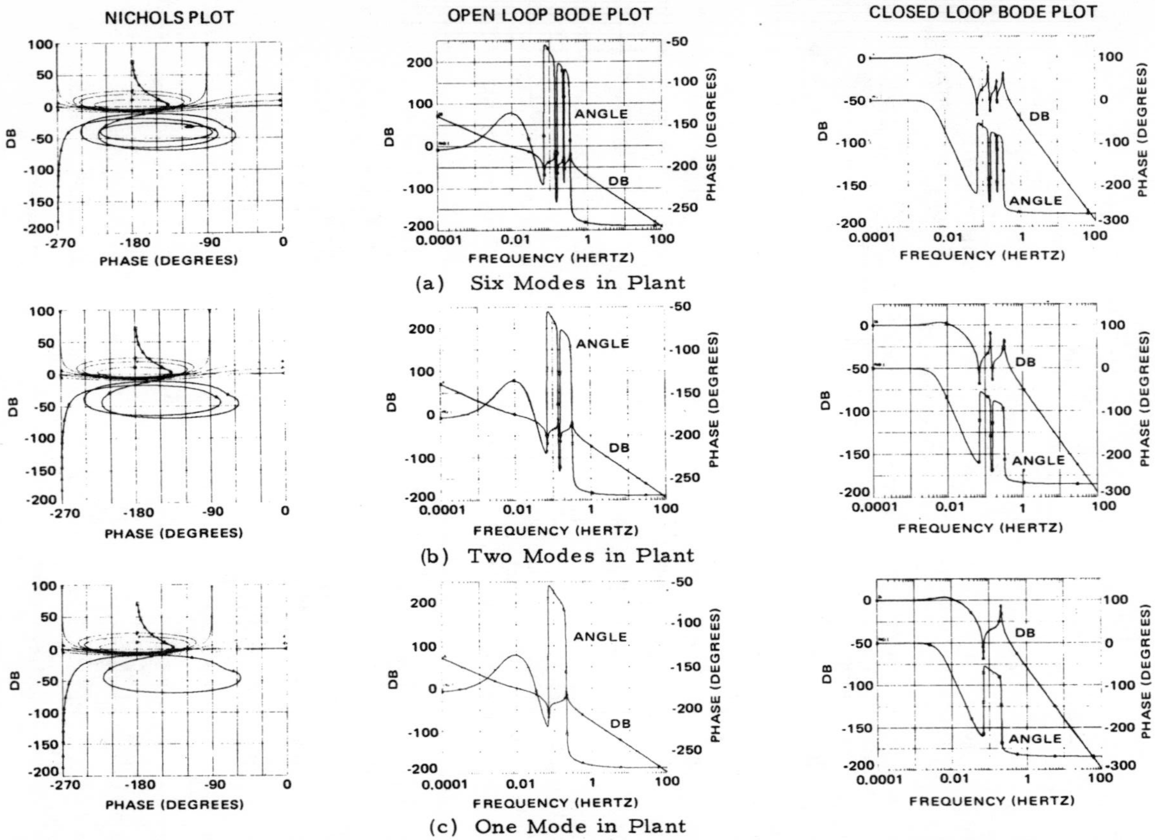

Fig. 12. Nichols, open-loop, and closed-loop Bode plots. Second-order filter with (a) six, (b) two, and (c) one mode(s) in plant.

figure as was drawn from Fig. 11.

Figure 13 shows the system Bode plots for the nominal parameter case with high-thrust engines. Note the greater gain margin for the sixth-order controllers. The gain margins for the sixth-, fourth-, and second-order filters are 16, 14, and 14 dB, respectively. Figure 14 shows the same results as Fig. 13 with the damping in the evaluation model changed to zero while using the nominal value of 0.005 in the filters. The gain margins for this case become 12.5, 8, and 8 dB for the sixth-, fourth-, and second-order filters, respectively.

VI. CONCLUSIONS

The techniques of modern control theory offer bright promise for practical applications such as spacecraft attitude control, but the efficacy of these methods in application to complex systems is seriously impaired by their reliance on good mathematical models of physical processes. Both because modeling capabilities are limited and because control system computers may be severely constrained in size, it is essential that we recognize that the mathematical model in the estimator (or filter) is not descriptive of the physical system (or plant) in all salient respects. This means that we must systematically explore the consequences of modeling errors in the estimator, perhaps following the rather straightforward practice of the present chapter in setting up various evaluation models against which to measure the consequences of errors in the estimator model. If these consequences are severe as indeed they are in some cases treated in this chapter, then we must be prepared to broaden the theory of modern controls to admit different models in the estimator and the plant, following the lead established in Refs. 10-13.

The examples treated here suggest that there are disturbingly unpredictable (or at least unpredicted) patterns in results obtained by playing with modeling errors; a seemingly rational

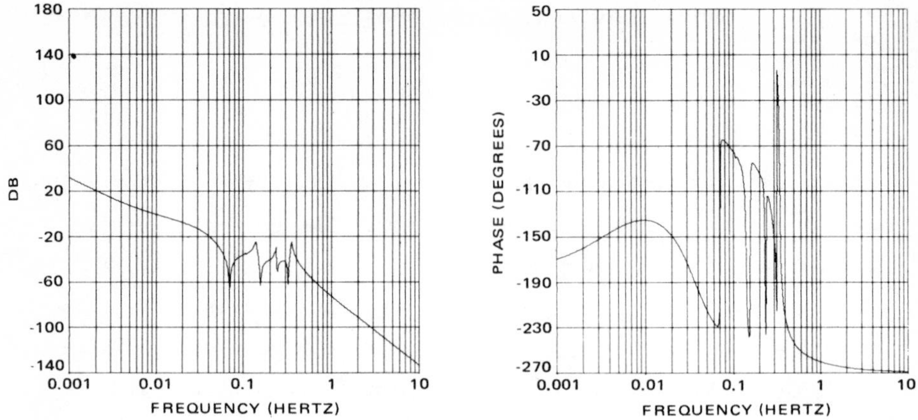

(a) Sixth Order Filter

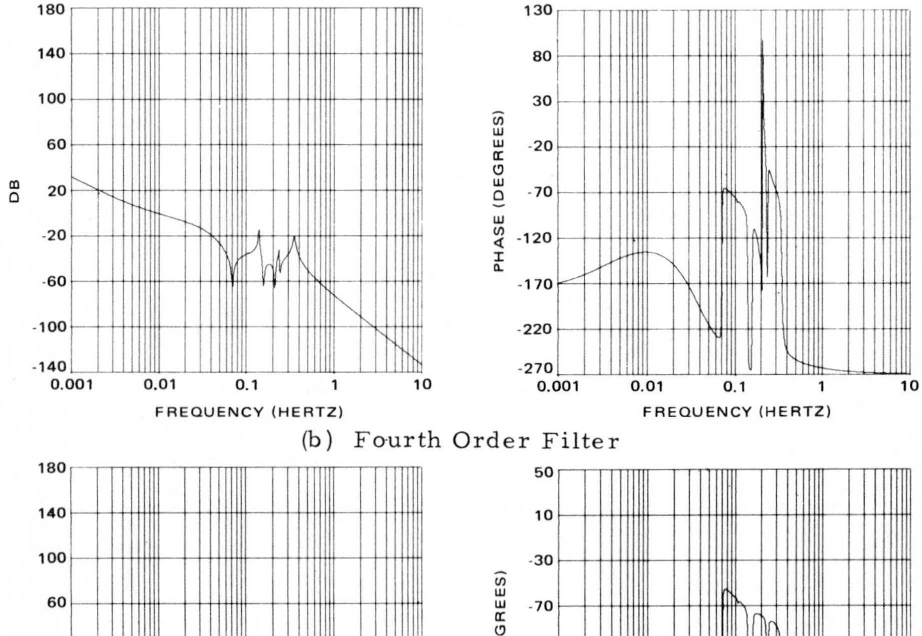

(b) Fourth Order Filter

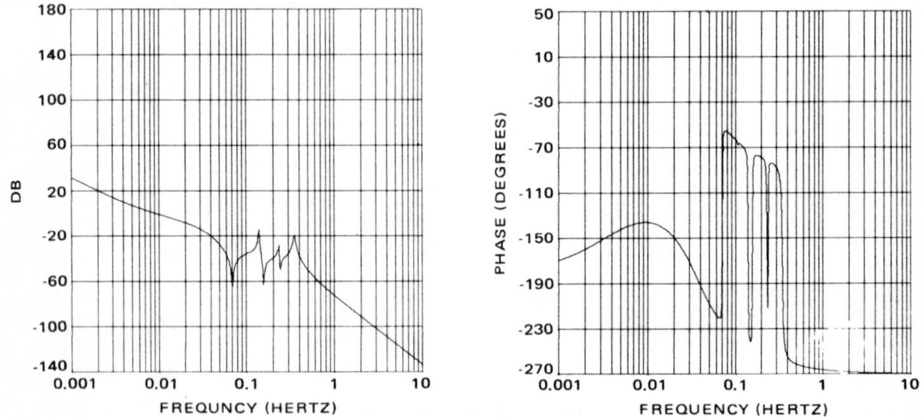

(c) Second Order Filter

Fig. 13. System open-loop Bode plots (six modes in plant, nominal parameters) for: (a) sixth-order; (b) fourth-order; and (c) second-order filters.

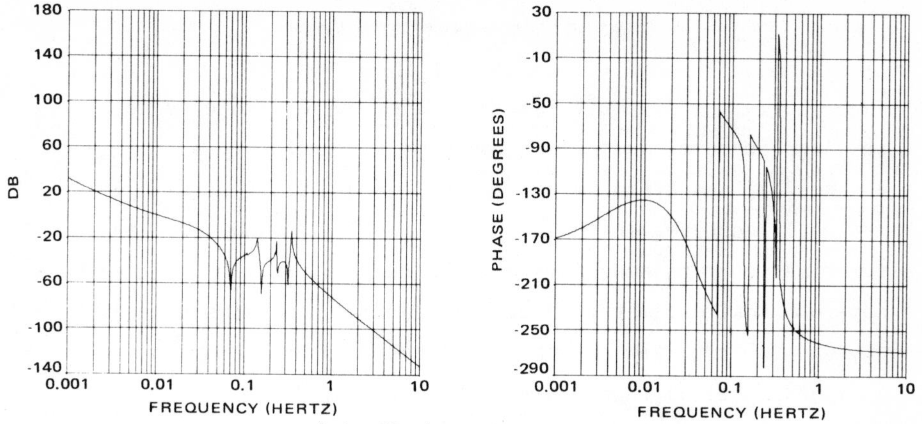

(a) Sixth Order Filter

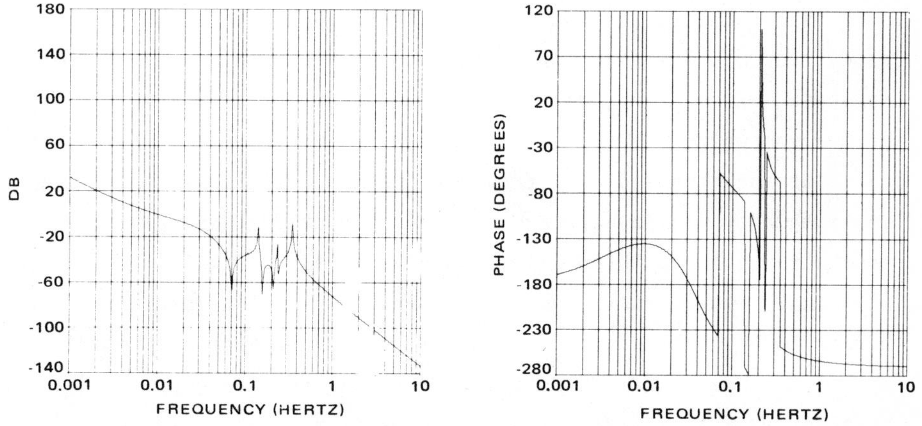

(b) Fourth Order Filter

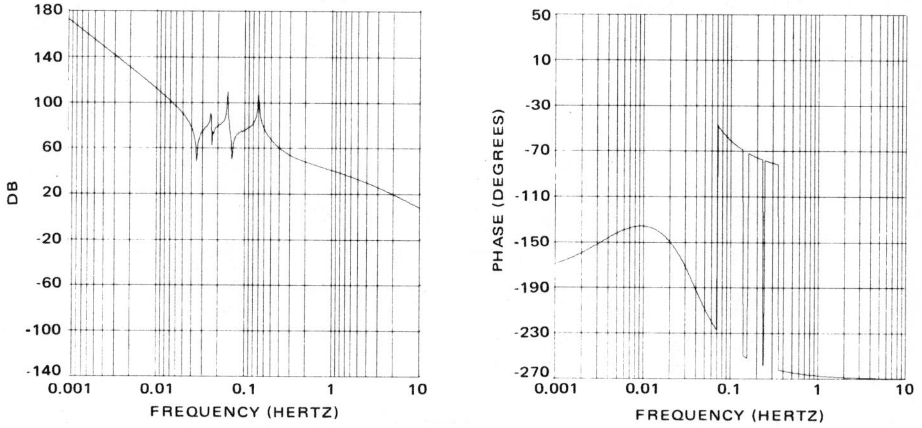

(c) Second Order Filter

Fig. 14. System open-loop Bode plots (six modes in plant, $\zeta = 0.005$ in filter) for (a) sixth-order, (b) fourth-order, and (c) second-order filters.

increase in the disparity between plant and estimator model resulted in some cases in improved performance (see again Figs. 8b and 7b), and attempts to improve performance by seemingly rational increases in the estimator model size and fidelity sometimes came to naught (see again Figs. 8d and 8b). It becomes apparent that estimator model size variation by simple model coordinate truncation as practiced here may not be optimum, and that new modeling theories must be considered. While we continue to believe that there are appropriate direct applications of existing theory, as considered in this chapter, we are also committed to the search for extensions of the theory which permit expansion of the range of practical applications of the methods of modern control theory.

The advantages of the frequency domain analysis are apparent. The frequency domain approach aided in the understanding of the modal truncation problem. It also allowed the similarity of classical notch filters and the higher-order filters presented herein to be observed. Additionally, using the frequency domain approach it was observed that an additional advantage of the sixth-order filter was that it had a greater gain margin than the lower-order filters.

REFERENCES

1. V.J. MODI, J. Spacecraft Rockets, 11, (11), 743-751 (1974).
2. V. LARSON, "A suboptimal stochastic controller for an N-body spacecraft", presented at the AAS/AIAA Astrodynamics Conf., Vail, Colorado, July 16-18, 1973; see also J. Spacecraft Rockets, 11, (12), 858-859 (1974).
3. D. POELAERT, A. LIÉGEOIS, and M. DUREIGNE, "Analysis and design of an attitude control system for nonrigid satellites", presented at the Euromech 39 Colloquium on Gyrodynamics, Louvain-La-Neuve, Belgium, September 3-5, 1973.

4. Y. OHKAMI and P. LIKINS, "The influence of flexibility on spacecraft controllability and observability", presented at the 6th Space Appl. Conf. Internat. Fed. Automatic Control, Tzakhadzor, USSR, August 1974; submitted for publication to Automatica.

5. L.S. SMITH and E.H. KOPF, "The development and demonstration of hybrid programmable attitude control electronics", presented at AGARD Conf. Real Time Computer Based Systems, Athens, Greece, May 27-31, 1974.

6. V. LARSON, P. LIKINS, and E. MARSH, "Optimal estimation and attitude control of a solar electric propulsion spacecraft," submitted to IEEE Trans. Aerospace Electronic Systems (1976).

7. A.E. BRYSON and Y.C. HO, "Applied Optimal Control". Ginn (Blaisdell), Boston, Massachusetts, 1969.

8. E.L. Marsh, "Multiple gimballing of thrusters for thrust vector control and thrust vector reorienting of solar electric spacecraft", presented at the AIAA 10th Electric Propulsion Conf., Lake Tahoe, Nevada, October 31-November 2, 1973; see also J. Spacecraft Rockets, 11, (11), 737-738 (1974).

9. E.L. MARSH, "The attitude control of a flexible solar electric spacecraft", presented at the AIAA 8th Electric Propulsion Conf., Stanford, California, August 31-September 2, 1970.

10. P. LIKINS, "Dynamics and control of flexible space vehicles", Jet Propulsion Laboratory Technical Report 32-1329, Rev. 1, January 15, 1970.

11. S.P. BHATTACHARYYA and J.B. PEARSON, S.W.I.E.E.E.C.O., San Antonio, Texas, 1970; also Internat. J. Control. 12, (5), 795-806 (1970).

12. C.D. JOHNSON, IEEE Trans. Automatic Control, AC-15, (2), 222-228 (1970); see also AC-15, (4), 516-518 (1970.

13. C.D. JOHNSON and R.E. SKELTON, AIAA J., 9, (1), 12-23 (1971).

14. R.E. SKELTON and P.W. LIKINS, "On the use of model error systems in the control of large scale linearized systems", submitted for presentation to the IFAC Symp. Large Scale Systems, to be held in Udine, Italy, June 1976.

15. Y. OHKAMI and P. LIKINS, "Eigenvalues and eigenvectors for hybrid coordinate equations of motion of flexible spacecraft", submitted to Celestial Mechanics.

16. V. LARSON, "An Analytical Stochastic Controller", Acta Astronautica, 2, 879-899 (1975).

STABILITY ANALYSIS OF STOCHASTIC INTERCONNECTED SYSTEMS

A. N. MICHEL AND R. D. RASMUSSEN

Department of Electrical Engineering
and Engineering Research Institute
Iowa State University
Ames, Iowa

† This work was supported by the National Science Foundation under Grant ENG 75-14093 and by the Engineering Research Institute, Iowa State University.

I. INTRODUCTION

The direct method of Lyapunov has found a wide range of applications in engineering and the physical sciences. Although most of these applications were originally concerned with the stability analysis of systems described by ordinary differential equations, the direct method of Lyapunov has also been extended to systems described by difference equations, integral equations, partial differential equations, differential difference equations, functional differential equations, stochastic differential equations, and the like. The stability theory of general dynamic systems is still of current interest.

Despite its elegance and generality, the usefulness of the direct method is severely limited when applied to problems of high dimension and intricate structure. For this reason it may be advantageous to view high-order systems as being composed of several lower-order systems which, when interconnected in an appropriate fashion, yield the original *composite* or *interconnected system*. The stability analysis of such systems can often be accomplished in terms of the simpler subsystems and in terms of the interconnecting structure of such composite systems. In this way, complications that usually arise when the direct method is applied to high-order systems may often be circumvented. This point of view has been adopted by several workers to analyze deterministic systems (see, e.g., Refs. [1-12]).

Even though disturbances are important considerations in the qualitative analysis of large-scale systems, apparently only a few reports [13-16] deal with the stability analysis of interconnected systems described by stochastic differential equations and stochastic difference equations. In the present chapter, stability results for several classes of continuous time parameter stochastic interconnected systems are presented. System disturbances described by Wiener processes, Poisson step processes, strong diffusion processes, finite state jump Markov processes, and the like are considered.

These disturbances may enter into the system description in both the subsystem structure and the interconnecting structure. Stability types considered include asymptotic stability and exponential stability with probability one, in probability, and in the quadratic mean. The objective is always the same: to analyze interconnected or composite systems in terms of their lower-order (and hopefully simpler) subsystems and in terms of their interconnecting structure. In order to demonstrate the usefulness of the present approach, several specific examples are presented.

In Section II, the notation employed throughout this chapter is established. The composite systems considered are formalized in Section III. In Section IV, the main stability results are presented, and then applied to several specific examples in Section V. The results of Section IV are proved in Section VI. Finally, the last section contains some concluding remarks.

II. NOTATION AND PRELIMINARIES

Let R^n denote Euclidean n-space, let $|\cdot|$ denote the Euclidean norm, let $\|\cdot\|$ denote an arbitrary norm, and let $x' = (x_1, \ldots, x_n)$ denote the transpose of $x \in R^n$. Let $\lambda(A)$ denote an eigenvalue of a square matrix A and let the norm of a rectangular matrix D, induced by the Euclidean norm, be denoted by $\|D\| = (\lambda_{\max}(D'D))^{1/2}$, where D' is the transpose of D. Also, for the rectangular matrix $D = ((d_{ij}))$, let $\|D\|_m = (\Sigma_{i,j} d_{ij}^2)^{1/2} = (\mathrm{tr}(D'D))^{1/2}$, where $\mathrm{tr}(A)$ denotes the trace of A.

Let $T = R^+ = [0, \infty)$ and let $\{x_t, t \in T\}$ denote a continuous parameter stochastic process of dimension n. Henceforth it is assumed that $x_0 = x$ is a known constant.

A function $\phi\colon R^+ \to R^+$ is said to be of <u>class</u> K (i.e., $\phi \in K$) if

(i) it is continuous over R^+,

(ii) $\phi(0) = 0$, and

(iii) $r_1 > r_2$ implies that $\phi(r_1) > \phi(r_2)$ for all $r_1, r_2 \in R^+$.

If $\phi \in K$ and $\lim_{r\to\infty} \phi(r) = \infty$, then ϕ is said to be *radially unbounded.*

Systems are considered which may be represented by equations of the form

$$dx = f(x, t)\, dt + \sigma(x, t)\, d\xi, \tag{1}$$

where $x \in R^n$, $t \in T$, $f\colon R^n \times T \to R^n$, $\sigma\colon R^n \times T \to R^{n\cdot m}$, and where $\{\xi_t, t \in T\}$ is an <u>independent increment Markov process</u>. In the present chapter two specific types of processes for Eq. (1) are considered: (a) <u>normalized m-dimensional Wiener processes</u> with independent components, denoted by $\xi_t = z_t$, and, (b) <u>normalized m-dimensional Poisson step processes</u> with independent components, denoted by $\xi_t = q_t$. It is assumed that $f(\cdot)$ and $\sigma(\cdot)$ satisfy conditions which ensure that for every $x_0 = x$ there exists a unique solution of Eq. (1)

with probability one (refer to Wong [17, p. 150] or Kushner [18, p. 13]). It is also assumed that $\{x_t = 0,\ t \in T\}$ is the only equilibrium of Eq. (1).

For the definitions of asymptotic stability with probability one (aswp1) and exponential stability with probability one (eswp1) of the equilibrium of Eq. (1), refer to Kushner [18, pp. 31-32]. For the definitions of asymptotic stability in probability (asip), exponential stability in probability (esip), and exponential stability in quadratic mean (esqm) of the equilibrium of Eq. (1), refer to Ia Kats and Krasovskii [19, pp. 1227 and 1234]. For additional good sources on stochastic stability, refer to Refs. [20-26].

Stability results for Eq. (1) involve the existence of Lyapunov-type functions $V: R^n \times T \to R^1$. Let $m > 0$ and define

$$Q_m = \{x \in R^n,\ t \in T: V(x, t) < m\}. \tag{2}$$

Henceforth it is assumed that all V-functions are such that

$\bigcup_{t \in R^+} \{x \in R^n: V(x, t) < m\}$ is a bounded subset of R^n.

Furthermore, it is assumed that $V(x, t)$ possesses continuous first- and second-order partials in x, and continuous first partials in t. Stability results for Eq. (1) also require that $V(x, t)$ be in the domain of the weak infinitesimal operator $\tilde{A}_{Q_m}$ for Eq. (1), given by

$$A_{Q_m} V(x, t) = \lim_{\delta \to 0^+} \frac{E_{x,t} V(x_{t+\delta}, t + \delta) - V(x, t)}{\delta},$$

$$(x, t) \in Q_m,$$

where $E_{x,t} V(x_{t+\delta}, t + \delta)$ denotes the conditional expectation of $V(x_{t+\delta}, t + \delta)$ given $x_t = x$ (see Refs. [17-21]).

If $\xi_t = z_t$, then A_{Q_m} is determined by

$$\tilde{A}_{Q_m} V = LV = \frac{\partial V}{\partial t} + \sum_i f_i(x, t) \frac{\partial V}{\partial x_i} + \frac{1}{2} \sum_{i,j} S_{ij}(x, t) \frac{\partial^2 V}{\partial x_i \, \partial x_j}$$

$$= \frac{\partial V}{\partial t} + f(x, t)'\nabla_x V(x, t) + \frac{1}{2} \sum_{i,j} S_{ij}(x, t)\nabla_{x_i x_j} V(x, t), \qquad (3)$$

where $S_{ij}(x, t)$ are defined by the matrix $((S_{ij}(x, t))) = \sigma(x, t)\sigma(x, t)'$.

If $\xi_t = q_t$ is a normalized Poisson step process, the independent components q_i of q experience a jump in any interval of length Δt with probability $\alpha_i \, \Delta t + o(\Delta t)$, where $o(\Delta t)$ denotes an infinitesimal of higher order than Δt. The amplitude distribution of q_i is given by $P_i(dq_i)$ and is such that

$$\int_{q_i} q_i P_i(dq_i) = 0, \qquad i = 1, \ldots, m$$

and

$$\int_{q_i} q_i P_i(dq_i) = \alpha_i^{-1}, \qquad i = 1, \ldots, m.$$

The corresponding infinitesimal generator is determined by

$$\tilde{A}_{Q_m} V = \mathcal{D}V = \frac{\partial V}{\partial t} + f(x, t)'\nabla_x V(x, t)$$

$$+ \sum_j \int [V(x + \sigma_j(x, t)\eta, t) - V(x, t)]\alpha_j \, dP_j(d\eta) \qquad (4$$

where $\sigma_i(x, t)$ denotes the ith column vector of $\sigma(x, t) = (\sigma_1(x, t), \ldots, \sigma_m(x, t))$.

Also considered are systems described by equations of the form

$$dx = f(x, t, \xi)\, dt, \tag{5}$$

were $x \in R^n$, $t \in T$, $\xi_t = y_t \in Y$ is a finite state jump Markov process, $f\colon R^n \times T \times Y \to R^n$ and where $\{x_t = 0, t \in T\}$ is assumed to be the only equilibrium of Eq. (5). The function $f(\cdot)$ is assumed to satisfy conditions that ensure the existence and uniqueness of the solution $\{x_t, t \in T\}$ (see Ia Kats and Krasovskii [19, pp. 1225, 1226]). For Eq. (5) one has (see Ia Kats and Krasovskii [19])

$$\tilde{A}_{Q_m} V = AV = \frac{\partial V}{\partial t} + \sum_i f_i(x, t, y^k) \frac{\partial V(x, t, y^k)}{\partial x_i}$$

$$+ \sum_i \alpha_{kj} [V(x, t, y^j) - V(x, t, y^k)]. \tag{6}$$

The constant α_{ij} in Eq. (6) is defined by

$$p_{ij}(\Delta t) = \alpha_{ij}\, \Delta t + o(\Delta t), \qquad i \neq j,$$

where $p_{ij}(\Delta t)$ denotes the probability of the change $y^i \to y^j$ during the time interval Δt.

III. COMPOSITE SYSTEMS

Composite or interconnected systems are considered that may be described by equations of the form

$$dw_i = f_i(w_i, t)\,dt + g_i(w_1, \ldots, w_\ell, t)\,dt$$

$$+ \sum_{j=1}^{\ell} (\sigma_{ij}(w_j, t)\,d\xi_j, \qquad i = 1, \ldots, \ell, \quad (7)$$

where $w_i \in R^{n_i}$, $f_i\colon R^{n_i} \times T \to R^{n_i}$, $\sigma_{ij}\colon R^{n_j} \times T \to R^{n_i \cdot m_j}$, $f_j(w_j, t) = 0$, $g_i(w_1, \ldots, w_\ell, t) = 0$, and $\sigma_{ij}(w_j\ t) = 0$ for all $t \in T$ if and only if $w_j = 0$, $\xi_i \in R^{m_i}$, ξ_{i_t} is an independent increment Markov process, and $g_i\colon R^{n_1} \times \cdots \times R^{n_\ell} \times T \to R^{n_i}$.

Letting $\sum_{j=1}^{\ell} n_j = n$, $\sum_{j=1}^{\ell} m_j = m$, $x' = (w_1', \ldots, w_\ell') \in R^n$, $f(x, t)' = [f_1(w_1, t)', \ldots, f_\ell(w_\ell, t)']$

$$g(x, t)' = [g_1(w_1, \ldots, w_\ell, t)', \ldots, g_\ell(w_1, \ldots, w_\ell, t)']$$

$$\triangleq [g_1(x, t)', \ldots, g_\ell(x, t)'],$$

letting $\sigma(x, t) = ((\sigma_{ij}\ (w_j, t)))$ and letting $\xi' = (\xi_1', \ldots, \xi_\ell')$, one can represent Eq. (7) equivalently as

$$dx = f(x, t)\,dt + \sigma(x, t)\,d\xi + g(x, t)\,dt$$

$$\triangleq F(x, t)\,dt + \sigma(x, t)\,d\xi, \quad (8)$$

where $f\colon R^n \times T \to R^n$, $\sigma\colon R^n \to R^{n \cdot m}$, $g\colon R^n \times T \to R^n$, and $\xi \in R^m$.

System (8), which is of the same form as Eq. (1), is called a composite system S (or an interconnected system S) having decomposition (7). It may be viewed as a nonlinear and time-varying interconnection (under disturbances) of ℓ isolated subsystems S_i of the form

$$dw_i = f_i(w_i, t)\, dt + \sigma_{ii}(w_i, t)\, d\xi_i. \tag{9}$$

The unique solutions of Eq. (9) are denoted by $\{w_{i_t}, t \in T\}$ with $w_{i_0} \triangleq w_i$.

Also considered are composite systems of the form

$$dw_i = f_i(w_i, t, \xi_i)\, dt + g_i(w_1, \ldots, w_\ell, t, \xi_1, \ldots, \xi_\ell)\, dt,$$
$$i = 1, \ldots, \ell, \tag{10}$$

where $w_i \in R^{n_i}$, $\xi_i \in Y_i \subset R^{m_i}$, ξ_{i_t} is a finite state jump Markov process, $f_i: R^{n_i} \times T \times Y_i \to R^{n_i}$ and $g_i: R^{n_1} \times \cdots \times R^{n_\ell} \times T \times Y_1 \times \cdots \times Y_\ell \to R^{n_i}$. Defining f, x, g, and ξ similarly as in Eq. (8), and letting $Y = Y_1 \times \cdots \times Y_\ell$, one can represent Eq. (10) equivalently as

$$dx = f(x, t, \xi)\, dt + g(x, t, \xi)\, dt \triangleq F(x, t, \xi)\, dt. \tag{11}$$

Equation (11) is clearly of the same form as Eq. (5). The isolated subsystems of composite system (11) are represented by

$$dw_i = f_i(w_i, t, \xi_i)\, dt, \qquad i = 1, \ldots, \ell. \tag{12}$$

In many practical cases, the following special interconnecting structure of composite systems (7) and (10) is of interest:

$$g_i(\cdot) = \sum_{j=1}^{\ell} g_{ij}(\cdot), \qquad i = 1, \ldots, \ell, \tag{13}$$

where $g_{ij}: R^{n_j} \times T \to R^{n_i}$ (resp., $g_{ij}: R^{n_j} \times T \times Y_1 \times \cdots \times Y_\ell \to R^{n_i}$

IV. MAIN RESULTS

For system (9) real-valued functions $V_i(\cdot)$ are employed with domain of definition in $R^{n_i} \times T$ and for system (12) real-valued functions $V_i(\cdot)$ are used with domain of definition in $R^{n_i} \times T \times Y_i$. Henceforth Eqs. (9) and (12) are referred to as "isolated subsystems S_i" and Eqs. (8) and (11) are called "composite" or "interconnected systems S."

DEFINITION 1. Isolated subsystem S_i possesses Property A if there exists a function $V_i(\cdot)$, two radially unbounded functions Ψ_{i1}, $\Psi_{i2} \in K$ and a function $\Psi_{i3} \in K$ such that

$$\Psi_{i1}(|w_i|) \leq V_i(\cdot) \leq \Psi_{i2}(|w_i|)$$

and

$$\tilde{A}_{Q_{m_i}} V_i(\cdot) \leq -\Psi_{i3}(|w_i|)$$

for all $(w_i, t) \in Q_{m_i}$, resp. $(w_i, t, y_i) \in Q_{m_i}$, and $m_i > 0$ arbitrarily large.

DEFINITION 2. Isolated subsystem S_i possesses Property B if there exists a function $V_i(\cdot)$ and three positive constants c_{i1}, c_{i2}, and c_{i3} such that

$$c_{i1}|w_i|^2 \le V_i(\cdot) \le c_{i2}|w_i|^2$$

and

$$\tilde{A}_{Q_{m_i}} V_i(\cdot) \le -c_{i3}|w_i|^2$$

for all $(w_i, t) \in Q_{m_i}$, resp. $(w_i, t, y_i) \in Q_{m_i}$, and $m_i > 0$ arbitrarily large.

DEFINITION 3. Isolated subsystem S_i is said to possess <u>Property C</u> if there exists a function $V_i(\cdot)$, two radially unbounded functions Ψ_{i1}, $\Psi_{i2} \in K$, and a function $\Psi_{i3} \in K$ such that

$$\Psi_{i1}(|w_i|) \le V_i(\cdot) \le \Psi_{i2}(|w_i|)$$

and

$$\tilde{A}_{Q_{m_i}} V_i(\cdot) \le \Psi_{i3}(|w_i|)$$

for all $(w_i, t) \in Q_{m_i}$, resp. $(w_i, t, y_i) \in Q_{m_i}$, and for $m_i > 0$ arbitrarily large.

DEFINITION 4. Isolated subsystem S_i is said to possess <u>Property D</u> if there exists a function $V_i(\cdot)$ and three positive constants c_{i1}, c_{i2}, and c_{i3} such that

$$c_{i1}|w_i|^2 \le V_i(\cdot) \le c_{iz}|w_i|^2$$

and

$$\tilde{A}_{Q_{m_i}} V_i(\cdot) \le c_{i3}|w_i|^2$$

for all $(w_i, t) \in Q_{m_i}$, resp. $(w_i, t, y_i) \in Q_{m_i}$, and for $m_i > 0$ arbitrarily large.

REMARK. If isolated subsystem S_i (described by Eqs. (9) and (12)) possesses Property A, then its equilibrium is (a) asymptotically stable in the large in probability (aslip) and (b) asymptotically stable in the large with probability one (aslwp1) (see Kushner [18] and Ia Kats and Krasovskii [19]). If isolated subsystem S_i (described by Eqs. (9) and (12)) possesses Property B, then its equilibrium is (a) exponentially stable in the large in probability (eslip), (b) exponentially stable in the large in quadratic mean (eslqm), and (c) exponentially stable in the large with probability one (eslwp1) (see Kushner [18] and Ia Kats and Krasovskii [19]). If isolated subsystem S_i possesses Property C or D, then its equilibrium may be unstable. For converse stability theorems, see Ia Kats and Krasovskii [19] and Kushner [26].

The proofs of the subsequent results are given in Section VI.

THEOREM 1. Assume that for composite system S (described by Eqs. (7) or (10)) the following conditions hold:

(i) In Eq. (7), $\{\xi_{i_t} = z_{i_t}, t \in T\}$ or $\{\xi_{i_t} = q_{i_t}, t \in T\}$, and $\sigma_{ij}(w_j, t) = 0$ for all $i \neq j$.

(ii) Let $M = \{1, \ldots, \ell\}$ and let $N \subset M$. Each isolated subsystem S_i (described by Eqs. (9) or (12), respectively) possesses Property A if $i \notin N$, $i \in M$, and Property C if $i \in N$.

(iii) For each scalar product $\nabla_{w_i} V_i(\cdot)' g_i(\cdot)$, $i = 1, \ldots, \ell$, there exist continuous real-valued functions $a_{ij}(x, t)$ resp. $a_{ij}(x, t, y)$ such that

$$\nabla_{w_i} V_i(\cdot)' g_i(\cdot) \leq [\Psi_{i3}(|w_i|)]^{1/2} \sum_{j=1}^{\ell} a_{ij}(\cdot) [\Psi_{j3}(|w_j|)]^{1/2}$$

holds for all $w_i \in R^{n_i}$, $w_j \in R^{n_j}$, $x \in R^n$, $t \in T$, resp. $y_i \in Y_i$.

(iv) There exists an ℓ-vector $\alpha' = (\alpha_1, \ldots, \alpha_\ell)$, $\alpha_i > 0$, $i = 1, \ldots, \ell$, and $\varepsilon > 0$ such that for each $x \in R^n$ and $t \in T$ resp. $y \in Y$, the matrix $S + \varepsilon I$ is negative definite, where I is the identity matrix and where $S = ((s_{ij}))$ is defined by

$$s_{ij} = \begin{cases} \alpha_i [a_{ii}(\cdot) - 1], & i \notin N \\ \text{and } \alpha_i [a_{ii}(\cdot) + 1], & i \in N, \quad i = j \\ \tfrac{1}{2}[\alpha_i a_{ij}(\cdot) + \alpha_j a_{ji}(\cdot)], & i \neq j \end{cases}$$

Then the equilibrium of composite system S (described by Eqs. (7) or (10)) is aslwpl and aslip.

THEOREM 2. Assume that for composite system (8) with decomposition (7) the following conditions hold:

(i) $\{\xi_{i_t} = z_{i_t}, \ t \in T\}$ (i.e., ξ_{i_t} is a normalized Wiener process with independent components) and assume that in general $\sigma_{ij}(w_j, t) \neq 0$, $i,j = 1, \ldots, \ell$.

(ii) Let $M = \{1, \ldots, \ell\}$ and let $N \subset M$. Each isolated system S_i described by Eq. (9) possesses Property C if $i \in N$ and Property A if $i \notin N$.

(iii) For each scalar product $\nabla_{w_i} V_i(w_i, t)' g_i(x, t)$ $i = 1, \ldots, \ell$, there exist continuous real-valued functions $a_{ij}(x, t)$ such that

$$\nabla_{w_i} V_i(w_i, t)' g_i(x, t) \leq [\Psi_{i3}(|w_i|)]^{1/2} \times \sum_{j=1}^{\ell} a_{ij}(x, t)[\Psi_{j3}(|w_j|)]^{1/2}$$

holds for all $w_i \in R^{n_i}$, $w_j \in R^{n_j}$, $x \in R^n$, and $t \in T$.

(iv) For each $V_i(w_i, t)$, $i = 1, \ldots, \ell$, there is a positive constant e_i such that

$$u_i'[\nabla_{w_i w_i} V_i(w_i, t)]u_i \leq e_i|u_i|^2$$

holds uniformly in w_i and $t \in T$ for all $u_i \in R^{n_i}$.

(v) For each $i,j = 1, \ldots, \ell$, $i \neq j$, there exists a continuous real-valued function $d_{ij}(w_j)$ such that $d_{ij}(w_j) \geq 0$ for all $w_j \in R^{n_j}$, and such that

$$\|\sigma_{ij}(w_j, t)\|_m^2 \leq d_{ij}(w_j)\Psi_{j3}(|w_j|)$$

for all $w_j \in R^{n_j}$ and $t \in T$.

(vi) There exists an ℓ-vector $\alpha' = (\alpha_1, \ldots, \alpha_\ell)$,

$\alpha_i > 0$, $i = 1, \ldots, \ell$, and $\varepsilon > 0$ such that for each $x \in R^n$ $t \in T$, the matrix $S + \varepsilon I$ is negative definite, where $S = ((s_{ij}))$ is defined by

$$s_{ij}(x, t) = \begin{cases} \alpha_i[a_{ii}(x, t) - 1] + \frac{1}{2}\sum\limits_{\substack{k=1,\\ i\neq k}}^{\ell} \alpha_k e_k d_{ki}(w_i), & i \notin N \\ & \qquad i=j \\ \alpha_i[a_{ii}(x, t) + 1] + \frac{1}{2}\sum\limits_{\substack{k=1,\\ i\neq k}}^{\ell} \alpha_k e_k d_{ki}(w_i), & i \in N \\ \frac{1}{2}[\alpha_i a_{ij}(x, t) + \alpha_j a_{ji}(x, t)], & i\neq j. \end{cases}$$

Then the equilibrium of composite system (8) is aslwpl and aslip.

THEOREM 3. Assume that for composite system S (described by Eqs. (7) or (10)) the following conditions hold:

(i) Hypotheses (i) and (ii) of Theorem 1 are satisfied.

(ii) For each scalar product $\nabla_{w_i} V_i(\cdot)'g(\cdot)$, $i = 1, \ldots, \ell$, there exist real constants a_{ij} such that

(a) $a_{ij} \geq 0$ for all $i \neq j$, and,

(b) $\nabla_{w_i} V_i(\cdot)'g_i(\cdot) \leq \sum\limits_{j=1}^{\ell} a_{ij}\, \Psi_{j3}(|w_j|)$

for all $x \in R^n$, $w_i \in R^{n_i}$, $w_j \in R^{n_j}$, $t \in T$ resp. $y_i \in Y_i$.

(iii) The matrix $S = ((s_{ij}))$ defined by

$$s_{ij} = \begin{cases} -a_{ii} + 1, & i \notin N \\ \text{and } -a_{ii} - 1, & i \in N, \end{cases} \quad i = j$$
$$\qquad -a_{ij}, \qquad\qquad\qquad i \neq j$$

has positive successive principal minors.

Then the equilibrium of composite system S (described by Eqs. (7) or (10)) is aslwpl and aslip.

THEOREM 4. Assume that for composite system (8) with decomposition (7) the following conditions hold:

(i) Hypotheses (i), (ii), and (iv) of Theorem 2 are satisfied.

(ii) For each scalar product $\nabla_{w_i} V_i(w_i, t)'g_i(x, t)$, $i = 1, \ldots, \ell$, there exist real constants a_{ij} such that

(a) $a_{ij} \geq 0$ for all $i \neq j$, and

(b) $\nabla_{w_i} V_i(w_i, t)'g_i(x, t) \leq \sum_{j=1}^{\ell} a_{ij}\Psi_{j3}(|w_j|)$

for all $x \in R^n$, $w_i \in R^{n_i}$, $w_j \in R^{n_j}$, $t \in T$.

(iii) Hypothesis (v) of Theorem 2 holds with $d_{ij}(w_j) \equiv d_{ij} \geq 0$.

(iv) The matrix $S = ((s_{ij}))$ defined by

$$s_{ij} = \begin{cases} -a_{ii} + 1, & i \notin N \\ \text{and} \quad -a_{ii} - 1, & i \in N, \quad i = j \\ -a_{ij} - \frac{1}{2} e_i d_{ij}, & i \neq j \end{cases}$$

has positive successive principal minors.

Then the equilibrium of composite system (8) is aslwpl and aslip.

THEOREM 5. Assume that for composite system S (described by Eqs. (7) or (10)) the following conditions hold:

(i) Hypothesis (i) of Theorem 1 holds.

(ii) The interconnecting structure of S is specified by Eq. (13).

(iii) Let $M = \{1, \ldots, \ell\}$ and let $N \subset M$. Each isolated subsystem S_i (described by Eqs. (9) or (12), respectively) possesses Property D if $i \in N$ and Property B if $i \notin N$, $i \in M$.

(iv) For each $i, j = 1, \ldots, \ell$, there is a constant b_{ij} such that

$$\nabla_{w_i} V_i(\cdot)' g_{ij}(\cdot) \leq b_{ij} |w_i| |w_j|$$

for all $w_i \in R^{n_i}$, $w_j \in R^{n_j}$, $t \in T$, resp. $y_i \in Y_i$.

(v) There exists an ℓ-vector $\alpha' = (\alpha_1, \ldots, \alpha_\ell)$, $\alpha_i > 0$, $i = 1, \ldots, \ell$, such that the matrix $S = ((s_{ij}))$ defined by

$$
s_{ij} = \begin{cases} \alpha_i(b_{ii} - c_{i3}), & i \notin N, \quad i \in M \\ \alpha_i(b_{ii} + c_{i3}), & i \in N, \quad i \in M \end{cases} \quad i = j
$$

$$
s_{ij} = \frac{1}{2}[\alpha_i b_{ij} + \alpha_j b_{ji}], \qquad i \neq j
$$

is negative definite.

Then the equilibrium of composite system S (described by Eqs. (7) or (10), respectively) is eslwpl, eslqm, and eslip.

THEOREM 6. Assume that for composite system (8) with decomposition (7) the following conditions hold:

(i) Hypotheses (i) and (iv) of Theorem 2 hold.

(ii) Hypothesis (iii) of Theorem 4 holds.

(iii) Hypotheses (ii), (iii), and (iv) of Theorem 5 hold.

(iv) There exists an ℓ-vector $\alpha' = (\alpha_1, \ldots, \alpha_\ell)$, $\alpha_i > 0$, $i = 1, \ldots, \ell$, such that the matrix $S = ((s_{ij}))$ defined by

$$
s_{ij} = \begin{cases} \alpha_i(b_{ii} - c_{i3}) + \frac{1}{2}\sum_{\substack{k=1,\\ i\neq k}}^{\ell} \alpha_k e_k d_{ki}, & i \notin N \\ \alpha_i(b_{ii} + c_{i3}) + \frac{1}{2}\sum_{\substack{k=1,\\ i\neq k}}^{\ell} \alpha_k e_k d_{ki}, & i \in N \end{cases} \quad i = j
$$

$$
s_{ij} = \frac{1}{2}[\alpha_i b_{ij} + \alpha_j b_{ji}], \qquad i \neq j
$$

is negative definite.

Then the equilibrium of composite system (8) is eslwpl, eslqm, and eslip.

REMARK. The inequalities involving bounds for $\nabla_{w_i} V_i(\cdot)' g_i(\cdot)$ in the hypotheses of the preceeding results are not particularly difficult to satisfy, nor do they restrict the form of the interconnections $g_i(\cdot)$ severely. For example, the bounds for $\nabla_{w_i} V_i(\cdot)' g_i(\cdot)$ in Theorems 1-4 are satisfied with $a_{ij}(x, t) \equiv a_{ij}$, a constant, if $f_i(\cdot)$ and $g_i(\cdot)$ are linear functions and if $V_i(\cdot)$ is quadratic.

REMARK. Theorems 1 and 3 (and likewise, Theorems 2 and 4) yield two different sets of sufficient conditions for the same stability concepts. In Theorems 3 and 4, the off-diagonal terms of the test matrices S are always nonpositive, which is not required in Theorems 1 and 2. On the other hand, the test matrices in Theorems 1 and 2 are always symmetric, which is not required in Theorems 3 and 4. In Theorems 1 and 2 it is possible to weigh the effects of isolated subsystems on total system performance via the weighting vector α, which is not the case in Theorems 3 and 4.

Experience with specific examples indicates that the sufficient conditions of Theorems 1 and 3 (and likewise, of Theorems 2 and 4) are not disjoint, i.e., for certain problems these theorems yield identical stability conditions. In certain cases, however, Theorem 1 yields less conservative results than Theorem 3 (likewise, Theorem 2 yields less conservative results than Theorem 4), and vice versa.

REMARK. The present results are applicable to composite systems with asymptotically stable subsystems and exponentially stable subsystems. Also, composite systems with unstable subsystems can be treated, provided that local negative feedback exists around each unstable subsystem, which is not an integral part of the corresponding isolated subsystem.

V. SOME EXAMPLES

Presently three specific examples are considered.

EXAMPLE 1. Consider the indirect control problem

$$\begin{aligned} dw_1 &= Aw_1\, dt + bf(w_2)\, dt + \sigma_{11}(w_1)\, dz_1 + \sigma_{12}(w_2)\, dz_2, \\ dw_2 &= [-\rho w_2 - rf(w_2)]\, dt + a'w_1\, dt + \sigma_{21}(w_1)\, dz_1 + \sigma_{22}(w_2)\, dz_2, \end{aligned} \tag{14}$$

where $w_1 \in R^{n_1}$, A is a stable $n_1 \times n_1$ matrix, b an n_1-vector, $r > 0$ a constant, a an n_1-vector, $\{z_{i_t},\ t \in T\}$ an m_i-dimensional normalized Wiener process, $\sigma_{ij}: R^{n_j} \to R^{n_i \cdot m_j}$, and $f(w_2)$ a real, single-valued function having the following properties: (i) $f(w_2)$ is continuous for all $-\infty < w_2 < \infty$, and (ii) $f(w_2) = 0$ if and only if $w_2 = 0$, and (iii) $0 < w_2 f(w_2) < kw_2^2$ for all w_2, where $k > 0$ is a constant.

Assume that for each $\sigma_{ij}(w_j)$ there is a constant $d_{ij} > 0$ such that $\|\sigma_{ij}(w_j)\|_m^2 \leq d_{ij}|w_j|^2$. Then, system (14), which is clearly a special case of composite system (8), may be viewed as a nonlinear interconnection under disturbances of two isolated subsystems:

$$S_1 :\ dw_1 = Aw_1\, dt + \sigma_{11}(w_1)\, dz_1, \tag{15}$$

$$S_2 :\ dw_2 = [-\rho w_2 - rf(w_2)]\, dt + \sigma_{22}(w_2)\, dz_2 \tag{16}$$

with interconnecting structure under disturbances specified by

$$g_{12}(w_2)\, dt + \sigma_{12}(w_2)\, dz_2 \triangleq bf(w_2)\, dt + \sigma_{12}(w_2)\, dz_2,$$

$$g_{21}(w_1)\, dt + \sigma_{21}(w_1)\, dz_1 \triangleq a'w_1\, dt + \sigma_{21}(w_1)\, dz_1,$$

where the notation of Eq. (7) is used.

Since A is stable there exists a positive definite matrix P such that the matrix $A^TP + PA \triangleq -Q$ is negative definite. Setting $V_1(w_1) = w_1'Pw_1$, and letting L_1 denote the operation L with respect to Eq. (15), one obtains

$$\nabla_{w_1} V_1(w_1)'g_{12}(w_2) = 2w_1'Pbf(w_2) \leq 2\lambda_{\max}(P)k|b|\,|w_1|\,|w_2|,$$

$$\nabla_{w_1w_1} V_1(w_1) = 2P,$$

$$\begin{aligned} L_1V_1(w_1) &= w_1'(A'P + PA)w_1 + \mathrm{tr}[\sigma_{11}(w_1)'P\sigma_{11}(w_1)] \\ &\leq [-\lambda_{\min}(Q) + \lambda_{\max}(P)\, d_{11}]\ |w_1|^2. \end{aligned}$$

Setting $V_2(w_2) \equiv w_2^2$, and letting L_2 denote the operation L with respect to (16), one has

$$\nabla_{w_2} V_2(w_2)'g_{21}(w_1) = 2w_2a'w_1 \leq 2|a|\,|w_1|\,|w_2|$$

$$\nabla_{w_2w_2} V_2(w_2) = 2,$$

$$\begin{aligned} L_2V_2(w_2) &= -2\rho w_2^2 - 2rw_2f(w_2) + [\sigma_{22}(w_2)]^2 \\ &\leq (-2\rho + d_{22})\ |w_2|^2. \end{aligned}$$

Choosing $\alpha_1 = 1/\lambda_{\max}(P)$ and $\alpha_2 = 1$, matrix S of Theorem 6 assumes the form

$$A = \begin{bmatrix} -\dfrac{\lambda_{\min}(Q)}{\lambda_{\max}(P)} + d_{11} + d_{21} & |b|k + |a| \\ |b|k + |a| & -2\rho + d_{22} + d_{12} \end{bmatrix}.$$

Matrix S is negative definite if

$$\rho > \frac{1}{2}(d_{12} + d_{22}) \tag{17}$$

and if

$$k < \frac{1}{|b|}\{[\frac{\lambda_{\min}(Q)}{\lambda_{\max}(P)} - d_{11} - d_{21}]^{1/2}[2\rho - d_{12} - d_{22}]^{1/2} - |a|\}. \tag{18}$$

Therefore, if inequalities (17) and (18) hold, then composite system (14) is eslwpl, eslqm, and eslip.

EXAMPLE 2. Consider the <u>system with two nonlinearities</u>

$$\begin{aligned} dw_i &= A_i w_i \, dt + b_i f_i(\theta_i) \, dt + \sigma_{i1}(w_1) \, dz_1 + \sigma_{i2}(w_2) \, dz_2, \\ \theta_i &= c_i' w_j, \qquad i = 1, 2, \qquad j = 1, 2, \qquad i \neq j, \end{aligned} \tag{19}$$

where $w_i \in R^{n_i}$, A_i is a stable $n_i \times n_i$ matrix, b_i an n_i-vector, c_1 an n_2-vector, c_2 an n_1-vector, $\{z_{i_t}, \ t \in T\}$ an m_i-dimensional normalized Wiener process, $\sigma_{ij}\colon R^{n_j} \to R^{n_i \cdot m_j}$, and $f_i(\theta_i)$ a single-valued continuous function with the properties $f_i(0) = 0$ and $|f_i(\theta_i)| \leq k_i|\theta_i|$

for all $\theta_i \neq 0$, where $k_i > 0$ is a constant.

Assume that for each $\sigma_{ij}(w_j)$ there is a constant $d_{ij} > 0$ such that

$$\|\sigma_{ij}(w_j)\|_m^2 \leq d_{ij}|w_j|^2 .$$

System (19), which is clearly a special case of composite system (8), may be viewed as a nonlinear interconnection under disturbances of two isolated subsystems

$$S_i: dw_i = A_i w_i \, dt + \sigma_{ii}(w_i) \, dz_i, \qquad i = 1, 2, \tag{20}$$

with interconnecting structure under disturbances specified by

$$g_{ij}(w_j) \, dt + \sigma_{ij}(w_j) \, dz_j = b_i f_i(c_i' w_j) \, dt + \sigma_{ij}(w_j) \, dz_j, \tag{21}$$

$$i = 1, 2, \qquad j = 1, 2, \qquad i \neq j.$$

Since A_i is stable there exists a positive definite matrix P_i such that $w_i'(A_i'P_i + P_iA_i)w_i \overset{\Delta}{=} -w_i'Q_iw_i$ is negative definite. Setting $V_i(w_i, t) \equiv V_i(w_i) = w_i'P_iw_i$, following the procedure of Example 1, and choosing $\alpha_i = 1/\lambda_{\max}(P_i)$, matrix S of Theorem 6 assumes the form

$$S = \begin{bmatrix} \dfrac{-\lambda_{\min}(Q_1)}{\lambda_{\max}(P_1)} + d_{11} + d_{21} & k_1|b_1||c_1| + k_2|b_2||c_2| \\ & \\ k_1|b_1||c_1| + k_2|b_2||c_2| & \dfrac{-\lambda_{\min}(Q_2)}{\lambda_{\max}(P_2)} + d_{22} + d_{12} \end{bmatrix}.$$

Matrix S is negative definite if

$$\frac{\lambda_{\min}(Q_1)}{\lambda_{\max}(P_1)} > d_{11} + d_{21} \tag{22}$$

and

$$k_1|b_1||c_1| + k_2|b_2||c_2| < \left[\frac{\lambda_{\min}(Q_1)}{\lambda_{\max}(P_1)} - d_{11} - d_{21}\right]^{1/2} \times \left[\frac{\lambda_{\min}(Q_2)}{\lambda_{\max}(P_2)} - d_{22} - d_{12}\right]^{1/2} . \tag{23}$$

Therefore, if inequalities (22) and (23) hold, then composite system (19) is eslwpl, eslqm, and eslip.

REMARK. When $V(\cdot) \equiv V(x)$ has a quadratic form, then $LV = \mathcal{D}V$ (see Eqs. (3) and (4)). Therefore, the conclusion of Examples 1 and 2 are also valid if in these examples the Wiener processes are replaced by Poisson step processes.

REMARK. The deterministic versions of Examples 1 and 2 were originally considered by Piontkovskii and Rutkovskaya, [2]. Their results were improved by Michel, [11]. Also, Examples 1 and 2 were treated by Michel [13] under the assumptions that $\sigma_{ij} = 0$ for all $i \neq j$ and assuming that the disturbances be Wiener processes.

EXAMPLE 3. Consider the system with multiplicative noise

$$\dot{x} = [A(x, t) + W(t)]x, \tag{24}$$

where $x \in R^{\ell}$, $t \in T$, A is an $\ell \times \ell$ array of continuous and bounded scalar functions $a_{ij}(x, t)$, and $W(t)$ is an array of wide-band Gaussian random processes $\bar{\sigma}_{ij} w_{ij}(t)$,

$i, j = 1, \ldots, \ell$. Each $w_{ij}(t)$ is assumed independent of the others with zero mean and unit variance. Assume also that $a_{ii}(x, t) = a_{ii}(x_i, t)$.

System (24) may be transformed (in the limit as each $w_{ij}(t)$ approaches "white noise") into an Ito differential equation. Letting $\sigma(x) = ((\sigma_{ij}(x_j)))$, where

$$\sigma_{ij}(x_j) = (0, \ldots, \bar{\sigma}_{ij}x_j, \ldots, 0), \qquad i, j = 1, \ldots, \ell, \tag{25}$$

(the nonzero term occuring in the jth position), and letting

$$v(t)' = (w_{11}(t), w_{21}(t), \ldots, w_{\ell 1}(t), w_{12}(t), \ldots, w_{\ell\ell}(t)), \tag{26}$$

Eq. (24) may be expressed alternatively as

$$\dot{x} = A(x, t)x + \sigma(x)v(t). \tag{27}$$

Following Wong [17, p. 162] one obtains the equivalent Ito differential equation

$$dx_i = [\sum_{j=1}^{\ell} a_{ij}(x, t)x_j + \frac{1}{2}\bar{\sigma}_{ii}^2 x_i]\, dt$$

$$+ \sum_{j=1}^{\ell} \sigma_{ij}(x_j)\, dz_j, \qquad i = 1, \ldots, \ell, \tag{28}$$

where $z_j \in R^{\ell}$, $j = 1, \ldots, \ell$.

From Eq. (25) it follows that

$$\|\sigma_{ij}(x_j)\|_m^2 = \bar{\sigma}_{ij}^2 x_j^2, \qquad i, j = 1, \ldots, \ell. \tag{29}$$

The isolated subsystems are chosen as

$$S_i\colon dx_i = [a_{ii}(x_i, t) + \frac{1}{2}\bar{\sigma}_{ii}^2]x_i\, dt + \sigma_{ii}(x_i)\, dz_i\,, \qquad i = 1, \ldots, \ell. \tag{30}$$

Choosing $V_i(x_i) = x_i^2/2$, then $e_i = 1$, $i = 1, \ldots, \ell$. Also

$$\begin{aligned} L_i V_i(x_i) &= (a_{ii}(x_i, t) + \frac{1}{2}\bar{\sigma}_{ii}^2)x_i^2 + \frac{1}{2}\operatorname{tr}[\sigma_{ii}(x_i)'\sigma_{ii}(x_i)] \\ &\le [\sup_{R\times T} a_{ii}(x_i, t) + \bar{\sigma}_{ii}^2]x_i^2, \qquad i = 1, \ldots, \ell \end{aligned} \tag{31}$$

and

$$\begin{aligned} \nabla_{x_i} V_i(x_i) a_{ij}(x, t) x_j &= x_i a_{ij}(x, t) x_j \\ &\le \sup_{R^\ell \times T} |a_{ij}(x, t)|\,|x_i|\,|x_j|, \\ &\qquad i, j = 1, \ldots, \ell\,. \end{aligned} \tag{32}$$

Choosing $\Psi_{i3}(r) = r^2$, the S matrix of Theorem 6 assumes the form

$$s_{ij} = \begin{cases} \alpha_i (\sup\limits_{R\times T} a_{ii}(x_i, t) + \bar{\sigma}_{ii}^2) + \frac{1}{2}\sum\limits_{\substack{k=1,\\ k\ne i}}^{\ell} \alpha_k \sigma_{ki}^2, & i = j \\ \frac{1}{2}[\alpha_i \sup\limits_{R^\ell\times T} |a_{ij}(x, t)| + \alpha_j \sup\limits_{R^\ell\times T} |a_{ji}(x, t)|], & i \ne j. \end{cases} \tag{33}$$

Therefore, if S (given by Eq. (33)) is negative definite for some choice of $\alpha_i > 0$, $i = 1, \ldots, \ell$, then system (28) (or, equivalently, system (24)) is eslwpl, eslqm, and eslip.

If, in particular, $\bar{\sigma}_{ij} = 0$ for all $i \neq j$, then it can be shown (see, e.g., Araki and Kondo [6]) that constants $\alpha_i > 0$, $i = 1, \ldots, \ell$, exist for which S is negative definite, provided there exist constants $\lambda_i > 0$, $i = 1, \ldots, \ell$ such that

$$\sup_{R \times T} a_{ii}(x_i, t) + \sum_{j=1}^{\ell} \frac{\lambda_j}{\lambda_i} \sup_{R^\ell \times T} \left| a_{ij}(x, t) \right| < -\bar{\sigma}_{ii}^2,$$

$$i = 1, \ldots, \ell. \tag{34}$$

Thus if $\sigma_{ij} = 0$, $i \neq j$ and if inequality (34) holds, then system (28) (or equivalently system (24)) is eslwpl, eslqm, and eslip.

REMARK. The deterministic version of Eq. (28) (or (24)) and its qualitative (i.e., stability) analysis arises in a great variety of disciplines: transistor circuits (see, e.g., Michel [12] and Sandberg [27]), economics (see, e.g., Metzler [28]), biology (see, e.g., May [29]), and the like.

VI. PROOF OF MAIN RESULTS

Before giving the proofs of the present results some preliminary remarks are in order.

A. Preliminaries.

(1) In each proof Lyapunov functions of the form

$$V(\cdot) = \sum_{i=1}^{\ell} \alpha_i V_i(\cdot) \tag{35}$$

are employed, where $\alpha_i > 0$, $i = 1, \ldots, \ell$. Henceforth, let $\alpha' = (\alpha_1, \ldots, \alpha_\ell)$.

(2) If in Eq. (7), $\{\xi_{i_t} = z_{i_t},\ t \in T\}$, then it is easily shown that

$$\tilde{A}V(x, t) = LV(x, t)$$

$$= \sum_{i=1}^{\ell} \alpha_i L_i V_i(w_i, t) + \sum_{i=1}^{\ell} \alpha_i g_i(x, t)' \nabla_{w_i} V_i(w_i, t)$$

$$+ \frac{1}{2} \sum_{\substack{i,j=1,\\ i \neq j}}^{\ell} \alpha_i \operatorname{tr}[\sigma_{ij}(w_j, t)' \nabla_{w_i w_i} V_i(w_i, t)$$

$$\times \sigma_{ij}(w_j, t)], \tag{36}$$

where $\operatorname{tr}[A]$ denotes the trace of a matrix A, and where L_i denotes the operation L specified by Eq. (3) for the ith isolated subsystem S_i characterized by Eq. (9).

(3) Choose $V(\cdot)$ as in Eq. (35) and let $Q_{m_i} = \{w_i \in R^{n_i}$, $t \in T$, resp. $y_i \in Y_i$: $V(\cdot) < m_i\}$. Let $m_i = m/\alpha_i$, $m > 0$. Then $\alpha_i V_i(\cdot) \leq \sum_{j=1}^{\ell} \alpha_j V_j(\cdot) = V(\cdot) < m$, which implies that $(w_i, t) \in Q_{m_i}$ (resp. $(w_i, t, y_i) \in Q_{m_i}$) whenever $(x, t) \in Q_m$ (resp. $(x, t, y) \in Q_m$).

(4) For composite system (7) with $\sigma_{ij}(w_j, t) \equiv 0$ for $i \neq j$, and for composite system (10) it is easily shown that for composite system S one has

$$\tilde{A}_{Q_m} V(\cdot) = \sum_{i=1}^{\ell} \alpha_i \tilde{A}_{Q_{m_i}} V_i(\cdot) + \sum_{i=1}^{\ell} \alpha_i g_i(\cdot)' \nabla_{w_i} V_i(\cdot), \tag{37}$$

where $\tilde{A}_{Q_{m_i}} V_i(\cdot)$ denotes the infinitesimal operator for subsystem S_i.

(5) In the proofs of Theorems 3 and 4 the following result is used.

THEOREM A. (see, e.g., Gantmacher [30] and Bellman [31]) If $A = ((a_{ij}))$ is a real $n \times n$ matrix such that $a_{ij} \leq 0$, $i \neq j$, and if all successive principal minors of A are positive, then A is nonsingular and all elements of A^{-1} are nonnegative.

B. Proof of Theorem 1.

Choose $V(\cdot)$ as in Eq. (35) and define Q_{m_i} as above. Then $(x, t) \in Q_m$ implies that $(w_i, t) \in Q_{m_i}$ (resp. $(x, t, y) \in Q_m$ implies that $(w_i, t, y_i) \in Q_{m_i}$), $i = 1, \ldots, \ell$. Since each subsystem S_i possesses Property A or Property C, it follows that

$$\Psi_1(|x|) \leq \sum_{i=1}^{\ell} \alpha_i \Psi_{i1}(|w_i|) \leq V(\cdot) \leq \sum_{i=1}^{\ell} \alpha_i \Psi_{i2}(|w_i|) \leq \Psi_2(|x|), \tag{38}$$

and thus, $V(\cdot)$ is positive definite, decrescent, and radially unbounded.

In view of hypotheses (i) - (iii) and in view of Eq. (37) one has

$$\tilde{A}_{Q_m} V(\cdot) \leq \sum_{\substack{i=1,\\ i\notin N}}^{\ell} -\alpha_i \Psi_{i3}(|w_i|) + \sum_{\substack{i=1,\\ i\in N}}^{\ell} \alpha_i \Psi_{i3}(|w_i|)$$

$$+ \sum_{i=1}^{\ell} \alpha_i \{[\Psi_{i3}(|w_i|)]^{1/2} \sum_{j=1}^{\ell} a_{ij}(\cdot)[\Psi_{j3}(|w_j|)]^{1/2}\}$$

$$= \sum_{\substack{i=1,\\ i\notin N}}^{\ell} [-\alpha_i + \alpha_i a_{ii}(\cdot)]\{[\Psi_{i3}(|w_i|)]^{1/2}\}^2$$

$$+ \sum_{\substack{i=1,\\ i\in N}}^{\ell} [\alpha_i + \alpha_i a_{ii}(\cdot)]\{[\Psi_{i3}(|w_i|)^{1/2}\}^2$$

$$+ \sum_{\substack{i,j=1,\\ i\neq j}}^{\ell} \alpha_i a_{ij}(\cdot)[\Psi_{i3}(|w_i|)]^{1/2}[\Psi_{j3}(|w_j|)]^{1/2} .$$

Letting $u' = ([\Psi_{13}(|w_1|)]^{1/2}, \ldots, [\Psi_{\ell 3}(|w_\ell|)]^{1/2})$, and letting $Q = ((q_{ij}))$, where

$$q_{ij} = \begin{cases} \alpha_i[a_{ii}(\cdot) - 1], & i \notin N \quad i = j \\ \alpha_i[a_{ii}(\cdot) + 1], & i \in N \\ \alpha_i a_{ij}(\cdot), & i \neq j \end{cases}$$

one has

$$\tilde{A}_{Q_m} V(\cdot) \leq u'Qu = u'[\tfrac{1}{2}(Q + Q')]u = u'Su,$$

where S is defined in hypothesis (iv). Using hypothesis (iv) one has

$$\tilde{A}_{Q_m} V(\cdot) \leq u'Su \leq -\varepsilon|u|^2 = -\varepsilon \sum_{i=1}^{\ell} \Psi_{i3}(|w_i|)$$

$$\leq -\varepsilon\Psi_3(|x|) \tag{39}$$

for all $x \in R^n$, $t \in T$, resp. $y \in Y$, and for arbitrarily large $m > 0$.

Noting that Ψ_1, $\Psi_2 \in K$ are radially unbounded and that $\Psi_3 \in K$, the conclusion of the theorem follows from inequalities (38) and (39).

C. Proof of Theorem 2.

Choosing $V(x, t)$ as in Eq. (35), it is clear that

$$\Psi_1(|x|) \leq \sum_{i=1}^{\ell} \alpha_i \Psi_{i1}(|w_i|)$$

$$\leq V(x, t) \leq \sum_{i=1}^{\ell} \alpha_i \Psi_{i2}(|w_i|) \leq \Psi_2(|x|) \tag{40}$$

for all $(x, t) \in Q_m$ and for $m > 0$ arbitrarily large. Since each S_i possesses either Property A or C, it follows that Ψ_1, $\Psi_2 \in K$ are radially unbounded.

In view of hypotheses (i)-(v) and in view of Eq. (36) one has

$$LV(x, t) = \sum_{i=1}^{\ell} \alpha_i L_i V_i(w_i, t) + \sum_{i=1}^{\ell} \alpha_i g_i(x, t)'\nabla_{w_i} V_i(w_i, t)$$

$$+ \frac{1}{2} \sum_{\substack{i,j=1,\\ i\neq j}}^{\ell} \alpha_i \operatorname{tr}[\sigma_{ij}(w_j, t)' \nabla_{w_i w_i} V_i(w_i, t) \sigma_{ij}(w_j, t)]$$

$$\leq \sum_{\substack{i=1,\\ i\notin N}}^{\ell} -\alpha_i \Psi_{i3}(|w_i|) + \sum_{\substack{i=1,\\ i\in N}}^{\ell} \alpha_i \Psi_{i3}(|w_i|)$$

$$+ \sum_{i,j=1}^{\ell} \alpha_i a_{ij}(x, t) [\Psi_{i3}(|w_i|)]^{1/2} [\Psi_{j3}(|w_j|)]^{1/2}$$

$$+ \frac{1}{2} \sum_{\substack{i,j=1\\ i\neq j}}^{\ell} \alpha_i e_i \|\sigma_{ij}(w_j, t)\|_m^2$$

$$\leq \sum_{\substack{i=1,\\ i\notin N}}^{\ell} \alpha_i [a_{ii}(x, t) - 1] \Psi_{i3}(|w_i|)$$

$$+ \sum_{\substack{i=1,\\ i\in N}}^{\ell} \alpha_i [a_{ii}(x, t) + 1] \Psi_{i3}(|w_i|)$$

$$+ \sum_{\substack{i,j=1,\\ i\neq j}}^{\ell} \alpha_i a_{ij}(x, t) [\Psi_{i3}(|w_i|)]^{1/2} [\Psi_{j3}(|w_j|)]^{1/2}$$

$$+ \frac{1}{2} \sum_{\substack{i,j=1,\\ i\neq j}}^{\ell} \alpha_i e_i d_{ij}(w_j) \Psi_{j3}(|w_j|)$$

$$= \sum_{\substack{i=1\\ i\notin N}}^{\ell} \{\alpha_i [a_{ii}(x, t) - 1] + \frac{1}{2} \sum_{\substack{k=1,\\ k\neq i}}^{\ell} \alpha_k e_k d_{ki}(w_i)\} \Psi_{i3}(|w_i|)$$

$$+ \sum_{\substack{i=1, \\ i \in N}}^{\ell} \{\alpha_i [a_{ii}(x, t) + 1] + \frac{1}{2} \sum_{\substack{k=1, \\ k \neq i}}^{\ell} \alpha_k e_k d_{ki}(w_i)\} \Psi_{i3}(|w_i|)$$

$$+ \sum_{\substack{i,j=1 \\ i \neq j}}^{\ell} \alpha_i a_{ij}(x, t) [\Psi_{i3}(|w_i|)]^{1/2} [\Psi_{j3}(|w_j|)]^{1/2} .$$

Letting $u' = ([\Psi_{13}(|w_1|)]^{1/2}, \ldots, [\Psi_{\ell 3}(|w_\ell|)]^{1/2})$ and $Q = ((q_{ij}))$, where

$$q_{ij} = \begin{cases} \alpha_i [a_{ii}(x, t) - 1] + \frac{1}{2} \sum\limits_{\substack{k=1, \\ k \neq i}}^{\ell} \alpha_k e_k d_{ki}(w_i), & i \notin N \\ & \qquad i = j \\ \alpha_i [a_{ii}(x, t) + 1] + \frac{1}{2} \sum\limits_{\substack{k=1 \\ k \neq i}}^{\ell} \alpha_k e_k d_{ki}(w_i), & i \in N \\ \alpha_i a_{ij}(x, t), & \qquad i \neq j \end{cases}$$

one obtains

$$LV(x, t) \leq u'Qu = u'[\tfrac{1}{2}(Q + Q')]u = u'Su \leq -\varepsilon |u|^2$$

or

$$LV(x, t) \leq -\varepsilon \sum_{i=1}^{\ell} \Psi_{i3}(|w_i|) \leq -\varepsilon \Psi_3(|x|) \tag{41}$$

for all $(x, t) \in Q_m$ and $m > 0$ arbitrarily large.

Noting that Ψ_1, $\Psi_2 \in K$ are radially unbounded and that $\Psi_3 \in K$, the conclusion of the theorem follows from inequalities (40) and (41).

D. Proof of Theorem 3

Choose $V(\cdot)$ as in Eq. (35) for arbitrary $\alpha = (\alpha_1, \ldots, \alpha_\ell)$ $\alpha_i > 0$, $i = 1, \ldots, \ell$. Since each subsystem S_i possesses Property A or C, there exist again two radially unbounded functions Ψ_1, $\Psi_2 \in K$ such that

$$\Psi_1(|x|) \leq V(\cdot) \leq \Psi_2(|x|) \tag{42}$$

for all $(x, t) \in Q_m$ resp. $(x, t, y) \in Q_m$ and $m > 0$ arbitrarily large.

In view of hypotheses (i) and (ii) and Eq. (37) one has

$$\tilde{A}_{Q_m} V(\cdot) = \sum_{i=1}^{\ell} \alpha_i \tilde{A}_{Q_{m_i}} V_i(\cdot) + \sum_{i=1}^{\ell} \alpha_i g_i(\cdot)' \nabla_{w_i} V_i(\cdot)$$

$$\leq \sum_{\substack{i=1,\\ i \notin N}}^{\ell} -\alpha_i \Psi_{i3}(|w_i|) + \sum_{\substack{i=1,\\ i \in N}}^{\ell} \alpha_i \Psi_{i3}(|w_i|)$$

$$+ \sum_{i,j=1}^{\ell} \alpha_i a_{ij} \Psi_{j3}(|w_j|).$$

Letting $u' = [\Psi_{13}(|w_1|), \ldots, \Psi_{\ell 3}(|w_\ell|)]$ and defining $S = ((s_{ij}))$ as in hypothesis (iii), one has

$$\tilde{A}_{Q_m} V(\cdot) \leq -\alpha' S u.$$

Let $y' = \alpha' S$. Then $\tilde{A}_{Q_m} V(\cdot) \leq y' u$. By hypotheses (ii) and (iii), $s_{ij} \leq 0$ for $i \neq j$ and S has positive successive principal minors for all $x \in R^n$. Therefore, S^{-1} exists for all $x \in R^n$ and all elements of S^{-1} are nonnegative, by Theorem A. Thus

$$\alpha = (S^{-1})'y.$$

Since each row and column of S^{-1} must contain at least one non-zero element one can always choose y, $y_i > 0$, $i = 1, \ldots, \ell$, so that $\alpha_i > 0$, $i = 1, \ldots, \ell$. Therefore

$$\begin{aligned} \tilde{A}_{Q_m} V(\cdot) &\le -y'u < 0, \qquad && x \ne 0 \\ &= 0, && x = 0, \end{aligned}$$

i.e., $\tilde{A}_{Q_m} V(\cdot)$ is negative definite. Therefore there exists a function $\Psi_3 \in K$ such that

$$\tilde{A}_{Q_m} V(\cdot) \le -\Psi_3(|x|) \tag{43}$$

for all $(x, t) \in Q_m$, resp. $(x, t, y) \in Q_m$ and $m > 0$ arbitrarily large.

The conclusion of the theorem follows now from inequalities (42) and (43).

E. Proof of Theorem 4.

Choose $V(x, t)$ as in Eq. (35) for arbitrary $\alpha' = (\alpha_1, \ldots, \alpha_\ell)$, $\alpha_i > 0$, $i = 1, \ldots, \ell$. By Hypothesis (i) there exist again radially unbounded functions Ψ_1, $\Psi_2 \in K$ such that

$$\Psi_1(|x|) \le V(x, t) \le \Psi_2(|x|) \tag{44}$$

for all $(x, t) \in Q_m$ and $m > 0$ arbitrarily large.

In view of hypotheses (i)-(iii) and Eq. (36) one has

$$LV(x, t) = \sum_{i=1}^{\ell} \alpha_i L_i V_i(w_i, t) + \sum_{i=1}^{\ell} \alpha_i g_i(w_i, t)' \nabla_{w_i} V_i(w_i, t)$$

$$+ \frac{1}{2} \sum_{\substack{i,j=1\\ i\neq j}}^{\ell} \alpha_i \operatorname{tr}[\sigma_{ij}(w_j, t)' \nabla_{w_i w_i} V_i(w_i, t) \sigma_{ij}(w_j, t)$$

$$\leq \sum_{\substack{i=1,\\ i\notin N}}^{\ell} -\alpha_i \Psi_{i3}(|w_i|) + \sum_{\substack{i=1,\\ i\in N}}^{\ell} \alpha_i \Psi_{i3}(|w_i|)$$

$$+ \sum_{i,j=1}^{\ell} \alpha_i a_{ij} \Psi_{j3}(|w_j|) + \frac{1}{2} \sum_{\substack{i,j=1,\\ i\neq j}}^{\ell} \alpha_i e_i \|\sigma_{ij}(w_j, t)\|_m^2$$

$$\leq \sum_{\substack{i=1,\\ i\notin N}}^{\ell} \alpha_i [a_{ii} - 1] \Psi_{i3}(|w_i|) + \sum_{\substack{i=1,\\ i\in N}}^{\ell} \alpha_i [a_{ii} + 1] \Psi_{i3}(|w_i|)$$

$$+ \sum_{\substack{i,j=1,\\ i\neq j}}^{\ell} \alpha_i [a_{ij} + \frac{1}{2} e_i d_{ij}] \Psi_{j3}(|w_j|).$$

Letting $u' = [\Psi_{13}(|w_1|), \ldots, \Psi_{\ell 3}(|w_\ell|)]$, defining $S = ((s_{ij}))$ as in hypothesis (iv), defining $y' = \alpha' S$, and using the identical argument as in the proof of Theorem 3, one has again

$$A_{Q_m} V(x, t) \leq -y'u < 0 \qquad x \neq 0$$

$$= 0, \qquad x = 0$$

and there exists a function $\Psi_3 \in K$ such that

$$A_{Q_m} V(x, t) \leq -\Psi_3(|x|) \tag{45}$$

for all $(x, t) \in Q_m$ and $m > 0$ arbitrarily large.

The conclusion of the theorem follows from inequalities (44) and (45).

F. Proofs of Theorems 5 and 6.

The proofs of these statements are similar to the proofs of Theorems 1 and 2 and are therefore omitted.

VII. CONCLUDING REMARKS

Stability results for several classes of stochastic composite or interconnected systems are presented in this chapter. Since, in the present method, systems are analyzed in terms of their lower-order subsystems and in terms of their interconnecting structure, it is often possible to circumvent difficulties that arise when the Lyapunov approach is applied to high-dimensional systems.

The classes of problems that are considered are sufficiently general to accommodate disturbances represented by Wiener processes, Poisson step processes, and finite-state Markov chains which may enter into the system at the subsystem level and into the interconnecting structure. Under appropriate conditions the results are applicable to systems with unstable subsystems. The present results yield sufficient conditions for asymptotic stability in the large and exponential stability in the large with probability one, in probability, and in the quadratic mean.

To demonstrate the usefulness of the present approach, several specific examples are considered. These include, (a) the indirect control problem, (b) a feedback system with two nonlinearities, and (c) a system of equations representing a variety of problems, such as economic systems, nonlinear transistor networks, biological systems, and the like.

Using the method of analysis advanced herein, it is also possible to establish stability results for discrete parameter stochastic composite systems. In addition, it is possible to establish instability criteria which are analogous to the present stability results.

REFERENCES

1. F.N. BAILEY, SIAM J. Control, 3, 443 (1965).
2. A.A. PIONTKOVSKII and L.D. RUTKOVSKAYA, Automat. Remote Control, 10, 1422 (1967).
3. W.E. THOMPSON, IEEE Trans. Automatic Control, AC 15, 504 (1970).
4. M. ARAKI, K. ANDO, and B. KONDO, IEEE Trans. Automatic Control, AC 16, 22 (1971).
5. A.N. MICHEL and D.W. PORTER, IEEE Trans. Automatic Control, AC 17, 222 (1972).
6. M. ARAKI and B. KONDO, IEEE Trans. Automatic Control, AC 17, 537 (1972).
7. V.M. MATROSOV, Automat. Remote Control, 33, 1458 (1972).
8. V.M. MATROSOV, Automat. Remote Control, 34, 1 (1973).
9. L.T. GRUJIC and D.D. SILJAK, IEEE Trans. Automatic Control, AC 18, 636 (1973).
10. D.W. PORTER and A.N. MICHEL, IEEE Trans. Automatic Control, AC 19, 422 (1974).
11. A.N. MICHEL, SIAM J. Control, 12, 554 (1974).
12. A.N. MICHEL, IEEE Trans. Circuits Systems, CS 22, 305 (1975).
13. A.N. MICHEL, IEEE Trans. Automatic Control, AC 20, 246 (1975).
14. A.N. MICHEL, Z. Angew. Math. Mech. 55, 93 (1975).
15. A.N. MICHEL and R.D. RASMUSSEN, Proc. 12th Allerton Conf. Circuit and System Theory, Urbana, Illinois, October 1974, pp. 77-86.

16. R.D. RASMUSSEN and A.N. MICHEL, Proc. 3rd Milwaukee Symp. Automat. Computation and Control, Milwaukee, Wisconsin, April 1975, pp. 161-166.

17. E. WONG, "Stochastic Processes in Information and Dynamical Systems", McGraw-Hill, New York, 1971.

18. H. KUSHNER, "Stochastic Stability and Control", Academic Press, New York, 1967.

19. I. IA KATS and N.N. KRASOVSKII, Prikl. Math. Mech., 24, 1225 (1960).

20. H. KUSHNER, "Introduction to Stochastic Control", Holt, New York, 1971.

21. L. ARNOLD, "Stochastic Differential Equations: Theory and Applications", Wiley, New York, 1974.

22. F. KOZIN, Automatika, 5, 95 (1969).

23. F. KOZIN, Stability of the linear stochastic system, Proc. Symp. Stochastic Dynamical Systems, 294, 87, 1972.

24. J.E. BERTRAM and P.E. SARACHIK, IRE Trans. Circuit Theory, 6, 260, (1959).

25. H. BUNKE, Z. Angew. Math. Mech. 43, 63 (1963).

26. H. KUSHNER, SIAM J. Control, 5, 228 (1967).

27. I.W. SANDBERG, Bell System Tech. J., 48, 35 (1969).

28. L. METZLER, Econometrica, 13, 277 (1945).

29. R.M. MAY, Math. Biosci. 12, 59 (1971).

30. F.R. GANTMACHER, "The Theory of Matrices", Vol. II. Chelsea, Bronx, New York, 1964.

31. R. BELLMAN, "Introduction to Matrix Analysis", McGraw-Hill, New York, 1967.

INDEX

A
B 7
C 8
D 9
E 0
F 1
G 2
H 3
I 4
J 5